THE LIBRARY
OF
THE UNIVERSITY
OF CALIFORNIA
LOS ANGELES

GIFT OF

David Freedman

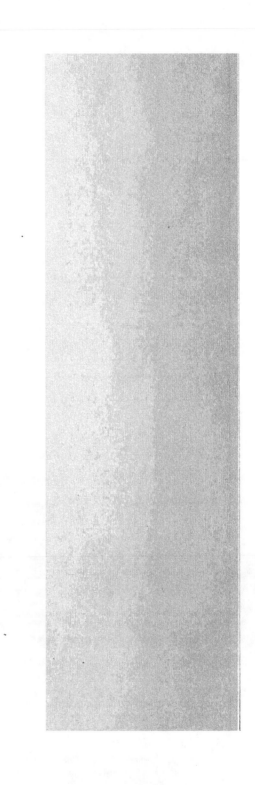

BRAN' NEW MONOLOGUES

Bran' New Monologues

And Readings in Prose and Verse

By
WALTER BEN HARE

Author of "The Boy Scouts," "The Camp Fire Girls," "A Couple of Million," "The Dutch Detective," "The Hoodoo," "Much Ado About Betty," "A Pageant of History," "Professor Pepp," "Teddy, or, the Runaways," "The Adventures of Grandpa," "The Scout Master," "On a Slow Train in Mizzoury," etc., and "Costume Monologues," and many others

BOSTON
WALTER H. BAKER & CO.
1923

Dedicated to

MISS *EMMA DEE RANDLE*

the *Chautauqua Star*

*who is wise enough to learn a
dozen new selections each season*

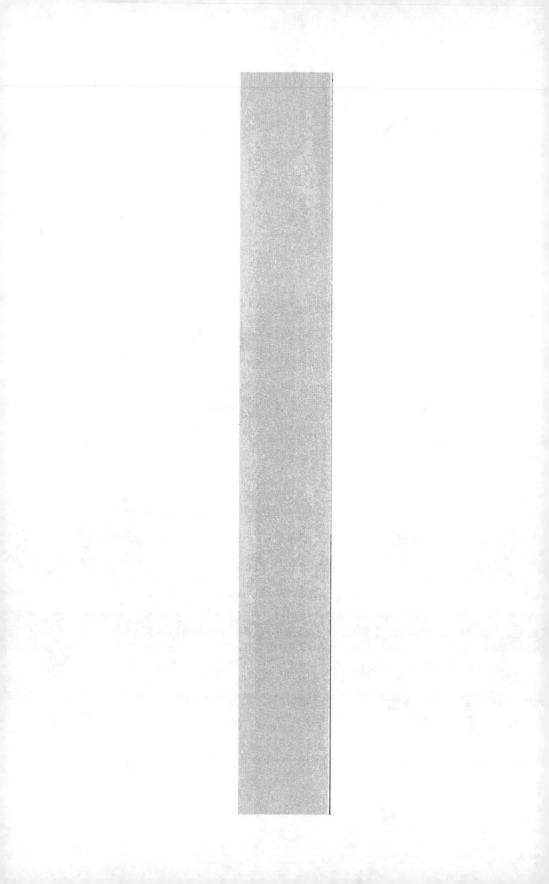

Contents

A LEAP YEAR LEAP—Comedy monologue for a débutante who decides to take advantage of leap year and propose to a bashful beau. A prize winner . 11

ON THE STREET CAR—Encore verse with a surprise ending 15

STELL AT THE PITCHER SHOW—Prose monologue, guaranteed to make any audience rock with laughter. As recited by the author with great success in all parts of the country. Released from royalty charges for the first time 16

TO JAMES WHITCOMB RILEY—Pathetic monologue introducing many bits from Riley's popular poems. A success everywhere. Just released from royalty charges 21

CHREESTOPHER COLUMBO—Comedy verse in Italian dialect 26

AN EXPRESSION-SCHOOL ROMANCE—Monologue for a stuttering girl. Refined comedy, full of bright lines and fun 29

ROMANCE—Encore bit 32

THE LITTLE BLACK CAT—Encore verse. Cat imitations. May be given with music if desired . . 33

FAWNCY—Encore verse, dude impersonation of a lisping man 34

IN DEFIANCE OF THE KAISER—Musical reading, dramatic and patriotic. A young American girl refuses to play the German national air . . . .

7

CONTENTS

"Uncle Tom's Cabin" at the Op'ry House—Musical monologue, with music inserted . . . 39

Bab's Birthday—Comedy monologue for a young wife, a selection that allows the reader to give the entire gamut of dramatic and comedy expression. Somewhat difficult but a decided hit . . . 42

Pastoral—Encore bit for a man, prose selection with a surprise ending 46

Sis Hopkins and Her Beau, Bilious—Stage monologue, country girl impersonation. A howling farce. Burlesque singing introduced. A never-failing success 47

Whoa, There, January!—Comedy story in prose of an old maid's race. Probably the most popular Yankee monologue ever published. From the repertoire of Emma Dee Randle 52

Deceitful Man—Encore selection in verse calling for impersonations of a child, a boy, a lover, a husband, a father and a grandfather. From the repertoire of Emma Dee Randle 58

Old King Faro's Daughter—Negro dialect monologue as recited by Prof. Vaughan of Memphis, Tenn. A never-failing success with all grades of audiences. Offered for the first time free of royalty charges 60

Sarah Jane—Musical monologue in verse, one simple selection to be played on the piano in several different ways. Very popular 66

Her First Ride in an Ottymobile—Character old-maid monologue with a laugh in every line and a clever, absorbing story. A Chautauqua favorite and a first prize winner in several contests. Just what the audiences like 69

Hiram's Blunder—Comedy encore verse . . 75

Things We See On the Stage—Protean monologue in verse describing various theatrical performances. Music introduced 76

CONTENTS

NORA HAS HER PICTURE TOOK—Irish comedy monologue. One of the most popular Irish selections ever written, a companion piece to "Nora and the Twins." As recited over 1,000 times by the author 80

THE MUSICALE—Prose encore bit calling for impersonation of several characters and music . . 86

A QUIET MAN AT A BASEBALL GAME—Comedy reading introducing the ever-popular baseball theme of excitement in the bleachers 88

DIFFERENT—Encore verse, very pleasing to children 91

MRS. GILHOOLY'S BUNGALOO—Irish dialect monologue. A side-splitting story 92

LILIAN—Prose encore selection, a "sell" on the audience. Popular 9

A CHRISTMAS HEROINE—Dialogue for two girls, an old maid and a slangy little waif. Very appealing. May be arranged as a Christmas reading . . 97

BOB'S GIRL—Encore verse, in character of a schoolboy of nine 104

TOMBOY—Comedy verse for a girl of eight or nine . 105

A Leap Year Leap

DEDICATED TO MISS MARTHA MOORE, THE MUSICAL
MONOLOGUIST

SCENE.—*A parlor. Character, a young society
girl in evening clothes.*

I've made up my mind at last. I'll throw convention to the winds and show the world that I'm a new woman. I'll do it—I'll do it to-night! I'll propose to Reginald Brady. (*A little faster.*) I've considered the question from every possible standpoint and I've come to the conclusion that a girl of to-day has just as much right to propose marriage as a man has. For centuries the poor girl has had to sit quietly by waiting for a man to snap his fingers before she can say, "Thank you!" Now, I'll do the snapping!

Reginald has been calling on me three times a week for the past year—but (*sadly*) never a word about love, never a thrill or a hand-clasp, never a syllable about matrimony. I can't stand it any longer, and I won't. To-night is my chance. I am to escort him to the Leap Year Ball at the Country Club and I intend to ask him right out, pointblank, marry me. Am I right? I'll say so.

in small pocket mirror.) I am, I know I am. My face is flushed, and I'm hot and my heart is beating like everything. But my mind is made up. To-night I'll make a leap year leap. I'll propose on the veranda overlooking the lake. (*Pulls a large chair forward.*) He'll be sitting there and I'll be sitting here. (*Pulls small chair close to large chair.*) No, I'll be closer than that. (*Puts small chair close to large chair.*) I'll start by resting my hand on the arm of his chair, like that. No, that isn't careless enough. That's better. Wait a minute, I'll have to have a Reginald. (*Places a pillow in large chair and puts a man's hat on it.*) Hello, Reggy! It's not a very striking likeness, but it's the best I can do.

Ah! (*Sentimentally*) isn't the moon bright to-night, Reginald? (*Pause.*) Don't you just adore a moonlight night? (*Pause.*) Yes, so do I. It makes me so sentimental. (*Pause.*) Don't you feel sentimental, too? The moon shining on the lake, and the music, and the perfume of the roses, and everything. (*With a long, audible sigh.*) Ohhh! it's just heavenly. (*Pause.*) Oh, no, I'm not the least bit chilly. Chilly? Why, I'm burning up. (*Sentimentally.*) 'Twas on such a night as this that what-cha-call her stood on the banks and waved a willow wand at Cypress. A night for romance, a night for love.

(*Matter-of-fact tone.*) That isn't very good. It doesn't seem to lead to anything. No, 's too much introduction. I'll start ri ̱t in   ritical ment. (*Deep, sepulchr* mal voice.*) Oh, no to death. (*High (Normal.*) T or something. Er—(*Clears throat*) hum! Reginald! That's much better. Reginald, the subject I am about to introduce will probably cause you some surprise. (*To audience.*) I should think it would. (*To dummy.*) But I trust it will cause you some feeling of joy. You surely must have learned during the past year—during the past year—you must have learned—that I—that you—that we—both of us—during the past year (*Clears throat*) hem! You must have learned—— (*To audience.*) Oh, fudge, I can't make it sound natural, at all.

(*In a confidential tone to the dummy.*) Say, Reggy, you and I seem to hit it off awfully well together. We've seen a lot of each other during the past year and we always get on like a house afire, you and I. I was just wondering why we couldn't always get on that way together, all through life, I mean, until death do us part. You know what I mean. (*To audience.*) That's splendid.

(*To dummy.*) I never cared for any other man the way I care for you, Reginald. Don't you care a little for me, too? If you do, why don't you say so, and make me the happiest——! (*Rises suddenly in alarm, as the maid is supposed to have entered the room.*) Who's there? Mercy, is that you, Marie? I wish you would knock before you enter my room. (*Pause.*) You *did* knock? Well, knock louder. I didn't hear you. I was just rehearsing a little scene from a play. What is it you wish? Please don't giggle. I must say, Marie, that you giggle more than any maid we ever had in the house. There's nothing at all to giggle at. Oh, you have a letter for me? A special delivery. Thank you. (You may go—and, Marie, you know that pink slip of mine you were admiring yesterday? I think I'll give it to you. (*Pause.*) Oh, never mind. You're welcome, I'm sure.

(*Pantomime looking at envelope.*) Reggy's writing! Oh, he must be sick or something. (*Opens imaginary letter and reads it.*) Oh! (*Read some more, pantomiming surprise.*) Well, I never! (*Read some more, registering delight; read a few lines aloud*)—have long loved you! (*Give a long audible sign and read some more.*)—marry me at once. Oh, it's a proposal. Reginald has proposed. (*Long sigh of relief.*) Thank goodness. (*Go toward entrance and call.*) Marie, Marie, get my wrap. No, I can't wait. I'll have to telephone. (*At door.*) Get Central for me, Marie, right away, and call up Mr. Brady. I have something very important to tell him over the 'phone. (*Ecstatically.*) Oh, Reginald! (*Exit.*)

berella? Acted jest like she owned the pitcher show. Don't some folks make you tired? I should smile! Oh, here comes the organist. Ain't he swell? *Look* at that chin and them eyes. He's the grandest player I ever heard. He kin play the sad and trembly fer the emotion jis' as good as the jazz fer the comedy. (*Pause.*)

Married? I don't believe it. Who's been kiddin' you that way? To that fresh thing in the ticket office? Well, wouldn't that scald you! Oh, look, some folks there on the aisle is gettin' up. Let's get their seats. (*Rises.*) Hurry up, Lil. Come on! (*Pantomimes passing the crowd as before.*) Excuse me! Say, will you please git up so'z a lady kin get past you? (*Working toward* C.) Boy, don't wipe your feet on *me!* I ain't no axminister. Hurry up, Lil, there's another couple's spotted the seats. (*Sits at* L.) No, you can't have that seat, it's engaged. Come on, Lil, I'm savin' yer seat fer y'. There's two good seats right over there. (*Points to* R.) We just left 'em, they're jist as good as these. (*To* LIL.) She must think she owns these seats.

Say, Lil, look over there quick. By the aisle. It's Mabel an' Jack. Wouldn't that freeze you! An' her jist the same as 'nagaged to Mr. Baker. Some'un had orter tell him. What *is* this pitcher, anyway? Who's that? It looks like Charley Ray, but where's his leading lady, Nassy Moover? He never plays with no other lead but her. They was married last month. (*Pause.*) It *ain't* Charley Ray? Huh, I guess I know Charley Ray when I see him. Didn't he play the part of the Miracle Man in Broken Blossoms? No, I guess it ain't, neither. This is a comic pitcher, an' Charley never plays in comics. It must be Henry Walthall. He's swell, actually the funniest guy on the screen. I nearly died at him in Doctor Jeckyl and Mr. Hyde. His wife's Mary Miles Minter—the one that plays the vamps in the Keystone Comedies. Sweetest thing! You orter see her play the snake-charmer in Cleopatra. It just worked me all up—the snake bites her in the seventh reel.

This ain't no comedy, after all. That feller wasn't the star at all. Here's the star; it's Wallace Reid. Oh, Lil, I just love his eyebrows, don't you? And the way his hair grows! And them shoulders! What they goin' to do? (*Pause, registering fright.*) Look, Lil, it's a cave er sump'm, and they're layin' fer Wallace. Three, four, five of 'em, and that villain Chinaman. They goin' to intice him there and beat him up, maybe kill him, so'z the Chinaman kin git the girl. Oh, look—Lil, there's gunpowder in them kegs. They're goin' to tie him to a keg and then set fire to the house. I seen it in the photographs outside. Look, they're hidin', and here he comes. Oh, I'm all trembly. They're hidin' under the steps and he's a-comin' down. Oh, that Chinaman's got a knife. (*Puts fingers over eyes.*) They're goin' to knife him! Tell me when they stab him, Lil. I can't bear to look at it. Wallace is so good-lookin'.

Is it over? Have they got him? (*Slowly takes her hands from eyes.*) Oh, look at the poor thing; he's bleedin'. They're draggin' him to that keg of gunpowder. I told you so. (*Suddenly.*) Oh, Lil, that Chinaman beat him in the face, and his hands all tied and everything. Ain't that brutal? (*Breathlessly.*) Now, they're goin' to fire the house. See, that's gasoline they're pourin' on the gunpowder to make it burn. Now they're leavin' him—and him all tied and helpless.

See that Chinese gal's goin' to tell the leadin' lady. She's goin' to save him. The Chinese gal's in love with him, too. Who could help it? Now she's at the leadin' lady's house. Oh, is *that* the leadin' lady? Wouldn't that frost you! She's readin' the message. She's goin' to save him. I don't think she's so much. She's got one of them last year's skirts, and look at her hair! I can't see how she got to be a leadin' lady. She ain't so many. She's gettin' on a horse. Look at her ride. She's got some class as a jockey, even if she is squint-eyed.

There's Wallace again. And the house is on fire. Oh, Lil, look at it burn. And him settin' on a keg of powder. The flames is creepin' closer and closer. Oh, oh! (*Pants with excitement.*) I'm so nervous! See! She's ridin' down the mountain. She's got to hurry! Go on, go on! There he is again. Look at the anguish that's writ on his face. Gee, I'll bet he's hot with all them flames so close to him. The floor's burnin'. Lil, the floor is burnin'. It's real fire, too. There she comes. Oh, looka, looka, she's beatin' in the door. Too late, the powder's goin' to ketch. Oh, oh! Gee, I swallered my gum!

She's saved him. (*Long sigh.*) She's pourin' water in his face to bring him to. Don't he look sad? Looka, he's comin' to. His clothes is all burnt, too—and looka the blood. Oh, Lil, he won't speak to her. Looka him turn his head away. He hates her, or sump'm, I wonder why, when she's the leadin' lady and saved his life and everything. Oh, he thinks she's married. Lil, he thinks she's married and has been making a plaything of his heart. What did it say? (*Reads slowly.*) Lovers Once but Strangers now! Ain't that sad! (*Pathetically.*) And he's jest breaking his heart fer her, too. (*Pause; her eyes fill with tears; she wipes them.*) Oh, why don't he see things right? She ain't married to the Chinaman at all. She jest thinks she is. Oh, he's goin' away. And she's too proud to tell him the truth. Lend me your handkerchief, Lil. (*Sniffs.*) Ain't it sad? I wonder if it's goin' to end like that with him goin' off with his heart broke and everything.

(*Brightens up.*) No, here comes the Chinese gal. She tells 'em it was all a lie and that the leadin' lady ain't married at all. Oh, look at him grab her. Ain't that swe-e-e-t! Say, that was a grand pitcher. But I can't give that leadin' lady a thing. Did you notice the way her skirt hung? Say, if she's a leadin' lady, well, you and me oughta be stars, that's all. She ain't got a thing on neither of us as fur as actin' goes, and as fer looks—huh! Well, come on, let's go. I ain't a-goin' to set through no comics. Oh, Lil, Mary Garden's comin' next week. She's just grand. She's the girl who took her name after the perfume. I saw her on the stage once in Bringin' Up Father, and she was the cutest thing. We'll have to come an' see her. Fatty Arbuckle's her leadin' man; you know him. He used to play heavies with Mary Pickford when she was starring in The Birth of a Nation. Come on, let's get some soda; I'm just dead. Whenever I see an emotional pitcher it jest kinda stirs me up, and I don't git over it fer days! Come on!

To James Whitcomb Riley

A Musical Monologue

I remember well the day I heard the news, jest like it might have been yesterday. Marthy was in the kitchen gettin' supper and Elsie was in the front room playin' on the piano. (*Sits at the piano.*) She was playin' my favorite tune, The Old Folks at Home. (*Plays chords on the piano.*) You know how it goes: (*Plays the chorus softly as she recites:*)

>All de world is sad and dreary,
>   Ev'rywhere I roam;
>Oh! darkies, how my heart grows weary,
>   Far from de old folks at home.

(*Plays verse and chorus softly as she recites:*) It was a summer day, jest the kind of a day *he* liked the best with "the chirrup of the robin and the whistle of the quail, as he piped across the meadows, sweet as any nightingale. When the bloom was on the clover and the blue was in the sky, and my happy heart brimmed over—in the days gone by." (*Ends music, but sits dreamily at the piano.*) I took up the weekly paper to see what was happenin' in my old home back in Indianny. I was readin' along, jest casual-like, you know, when I saw it. (*Plays same chords as before.*) It was jest a little newspaper notice, only about five inches of printed type, but when I read it the sun seemed to fade out of the sky, the birds all hushed their singin', and the tune Elsie was playin' in the front room seemed to change into a dirge. The whole world seemed changed, as if the Angel of Death had touched the face of summer and long streamers of black crepe hung on all the doors of men. James Whitcomb Riley was dead. (*End music.*)

I couldn't jest sense the meanin' of it at first, and I read that little notice over and over agin, thinkin' of him, our Jim! After a while I kinda got myself together-like and called the gals and told 'em. Well, sir, it was just like one of the fam'ly passing away, though none of us had ever laid eyes on Jim in our lives, but we knew him, we did—and he knew us. Why, the gals had been brought up on Riley's teachin', an' he was always at my right hand ready to help me through the rough paths, always ready with a cheerful word of advice, tellin' us that we should " be contented with our lot. The June is here this morning and the sun is shining hot. Oh! let us fill our hearts up with the glory of the day and banish every doubt and care and sorrow fur away! Fer the world is full of roses, and the roses full of dew, and the dew is full of heavenly love that drips fer me and you."

That's the kind of a man I like—big-hearted, lovin', faithful Jim Whitcomb Riley. I'll bet " there never was, on top of dirt, a feller better'n Jim! You want a favor, and couldn't git it anywheres else—you could git it from him. Give up every nickel he's worth; ef you'd a wanted it he'd give you the earth! Allus a-reachin' out, Jim was, and a-helpin' some poor fellow onto his feet. He never cared how hungry he was hisself, so's the *feller* got somethin' to eat. Favorite of the whole blamed neighborhood; when God made Jim, I'll bet He didn't do anything else that day but jest set around and feel good!"

I don't think he was what you'd call an educated man. He himself wrote that from childhood up tell he was old enough to vote he allus writ more er less poetry, and he writ it from the heart out. He said, "Thar is times when I write the tears roll down my cheeks." And didn't he know country life and country ways! There was his old swimmin' hole "whar the crick so still and deep looks like a baby-river that was lying half asleep."

And how he loved the out-a-doors, and sung about it, too: "When the frost is on the punkin and the fodder's in the shock, and you hear the kyouck and gobble of the struttin' turkey cock. Oh, it's then's the time a feller is a-feelin' at his best, with the risin' sun to greet him from a night of peaceful rest." He sings of "Spring, when the green gits back in the trees, and the sun comes out and *stays*, and yer boots pulls on with a good tight squeeze, and you think of your barefoot days." And of "wortermelon time is a-comin' 'round agin, and there ain't no man a livin' any tickelder'n me, fer the way I hanker arter wortermelon is a sin, which is the why and wherefore as you kin plainly see. I joy in my heart jes' to hear that rippin' sound when you split one down the back and jolt the halves in two, and the friends you love the best is gethered all around, and you says unto your sweetheart, 'Oh, here's the core fer you!'" And his message of inspiration: "There is ever a song somewhere, my dear; there is ever a something sings alway. There's the song of the lark when the skies are clear, and the song of the thrush when the skies are gray. The buds may blow and the fruit may grow, and the autumn leaves drop crisp and sear, but whether the sun, or the rain, or the snow, there is ever a song somewhere, my dear."

And how the children loved him, and how they'll miss him. Seems like no poetry-writer ever got so close to the real, livin', kickin', squirmin', inquisitive boy and gal like he did. No wonder they called him the Children's Poet. "He loved to hear their voices and wove them into his rhyme; and the music of their laughter was with him all the time. Though he knew the tongues of nations, and their meanings all were dear, the prattle and lisp of a little child was the sweetest for him to hear." (*Recite with music.*)

'Twas a funny little fellow
  Of the very purest type,
For he had a heart as mellow
  As an apple overripe;
And the brightest little twinkle
  When a funny thing occurred,
And the lightest little tinkle
  Of a laugh you ever heard.
He laughed away the sorrow
  And he laughed away the gloom
We are all so prone to borrow
  From the darkness of the tomb.
And he laughed across the ocean
  Of a happy laugh, and passed,
With a laugh of glad emotion
  Into Paradise at last.
And I think the angels knew him,
  And had gathered to await

An Expression-School Romance

Our heroine is a sweet young thing of some eighteen summers, a beautiful girl, a little giddy perhaps, but probably that is only the exuberance of youth. She is a student at a celebrated school of expression, not that she desires to become a public entertainer, or a parlor reader, or an actress, mercy, no!—she's there for medical purposes only, as she tells her chums, for Gwendolyn stutters, not an ordinary stammer, hem, pause, and go on—but a genuine, d-d-downright st-tut-tut-tutter. She has been a pupil in the school for three months, and all her friends have n-n-noticed a m-m-most wonderful improvement. She meets two of them on the campus of the school.

Oh, g-g-gur-girls! I've m-m-met him and he's s-s-(*Whistles*) simply the grandest man I ever s-s-s-(*Whistles*) saw. Eyes like Charles Ray and sh-sh-shoulders like D-d-doug. He's a new pupil. And I've j-just been introduced to him. His name is S-S-S (*Whistles*) Sampson. (*Exasperated.*) Oh, I'm stuttering again! After t-t-tw-tw-twenty-five lessons, t-t-too. (*Very slowly and distinctly.*) Whenever you are inclined to stutter think slowly, speak slowly, enunciate clearly. (*Faster.*) Whenever you are inclined to stutter think slowly, speak slowly, enunciate clearly. (*Rapidly.*) Whenever you are inclined to stutter think slowly, speak slowly, enunciate clearly. There! I never st-stutter when I say that, but I f-f-find it very trying to w-work that sentence into an ordinary conversation.

Oh, Sallie, he's the l-l-lu-loveliest thing! And he's taking p-private l-lessons with Madame. No c-c-c-(*Whistles*) class lessons at all. He w-went into her p-private room and he was all d-d-dressed in brown, and with his brown hair and brown eyes. Oh, j-j-gee! Wait a minute. (*Pantomimes putting beans in her mouth.*) Now I can talk like a human being. What is it? Beans, of course. That's the way I always did at home before I came to the school, but it's dreadfully inconvenient to carry beans in your mouth all the time. And Madame positively will not permit it.

His name is Clarence, and he's a college man. I think he's going on the stage, but I'm not sure. He'd make a wonderful leading man. He's quiet, though. He scarcely said a word when we were introduced, and I was afraid I'd stutter, so I just mumbled. (*Mumbles.*) Madame said he was much impressed with my reading last year at Commencement. He said it was the most natural thing he ever heard. I gave The Man Who Stuttered and the Girl Who Lisped, and he thought it was a wonderful impersonation. He didn't know that it was the real thing. Madame told me all this yesterday, and this morning when I saw him go in for his private lesson I thought it would be a good time to meet him.

How did I do it? Why, I went right in. I told Madame I wanted to borrow her Chart of Articulation, and he was standing right there, so she had to introduce us. His name is Morton, Clarence Morton. Isn't that a lovely name? He's from Philadelphia, and he has the dreamiest eyes. Shhh!

AN EXPRESSION-SCHOOL ROMANCE

Here comes Madame! (*Throws beans away.*) She d-d-doesn't allow b-b-b (*Whistles*) beans. I wonder if the lesson is f-f-f (*Whistles*) finished. Oh, she's going out. I thought she was coming here, and now I've l-l-lost my b-b-b-b (*Whistles*) beans.

Oh, S-Sallie, there he comes. He's c-c-coming this way. He's g-g-g-going to talk to m-m-me and I haven't any b-b-b (*Whistles*) beans. Oh, give me something quick. Pebbles, pins, anything! Candy? The very thing. These look j-j-just like b-b-b (*Whistles*) beans. (*Puts candy in her mouth.*) Don't run away. Sallie! You stay right here, I'll introduce you. (*Slight pause.*) They've gone. (*Turns and says sweetly.*) Oh, Mr. Morton, have you finished for the day? Don't hurry so. The campus is for all the students. Isn't it lovely? Won't you sit down? (*Sits.*) This is my favorite spot, here under this tree. I just love trees, don't you?

(*To audience.*) I wish he'd talk. He just nods or shakes his head.

I know! I'll make him say something! (*To him.*) I beg your pardon, but I didn't quite catch your name when we were introduced. Is it Morton or Wharton? (*Pause.*) I asked you if your name was Morton or Wharton. (*To audience.*) He doesn't talk—great heavens, maybe he can't talk. He's deaf and dumb. I wonder if I can make him understand. (*Talks to him with fingers.*) How wild he looks. He must think I'm c-c-crazy.

(*To audience.*) Oh, my candy beans; I've sw-sw-swallowed them. (*To him.*) I-I-I-I've swallowed my beans. I always st-t-tut-(*Whistles*) ter, when I haven't any beans in my m-m-mouth. (*Pause; looks at him in astonishment.*) Oh! You s-stut-stutter, too? That's the r-reason you wouldn't t-t-t (*Whistles*) talk? You didn't w-w-want me to know! How absurd. Why, I s-s-s (*Whistles*) sometimes s-st-stut-stutter a little myself. D-d-did you notice it? But I've t-t-t (*Whistles*) taken three months of lessons f-from M-M-Mad-DAM! Madame! So n-no one can h-hardly n-no-no-(*Whistles*) notice it at all.

(NOTE.—Real beans may be used during the first part of the monologue.)

Romance

They were sitting side by side—
   And he sighed and she sighed.
Said he, " My darling idol "—
   And he idled and she idled.
Said he, " Your hand I asked,
   So bold I've grown "—
And he groaned and she groaned.
Said she, " My dearest Luke! "—
   And he looked and she looked.
Said he, " Upon my heart there's such a weight! "
   And he waited and she waited.
Said he, " I'll have thee, if thou wilt "—
   And he wilted and she wilted.

The Little Black Cat

A little black cat sat under a tree,
    Me-ow, pur-r-rrr!
And sang a little song contentedly;
    Me-ow, me-ow,
    I'm happy now;
    Purr, pur-r-rrr!
    Thank you, sir!

A man and a maid came strolling by—
The maid was quizzing him, he was shy,
And the little black cat was wondering why!
    Me-ow, me-ow,
    There's going to be a row;
    Purr, pur-r-rrr!
    Look out, young sir!

The girl, a stern young Catechist,
Was putting the man through her question list.
"Now don't you know it was wrong?" she said.
The man shook a stubborn curly head.
"I don't see why a kiss is wrong,
    Of course I haven't known you long;
    But think of all the time I've missed;
    I'll make up now, Fair Catechist!"

And the little black cat down under the tree
Looked up at the pair apprehensively.

The rest of the scene I believe I missed,
But some one there was softly kissed—
It must have been the Catechist,
It surely wasn't the cat he kissed.
It's strange when you come to think of that,
He kissed the Catechist, not the cat.

And the little black cat sat under the tree,
And sang her little song contentedly:
      Purr, pur-r-rrr!
      It isn't a fight!
      Me-ow, me-ow,
      Kiss! Good-night!

NOTE.—The word kiss in the last line is not spoken by the reader, but the sound of a kiss is given instead of the word. The words "Good-night!" are given in the tone of an emphatic interjection with strong emphasis on each syllable.

Fawncy!

(Impersonate a lisping dude.)

Alith and I went walking ovah in Bothton town,
I in me long Pwinth Albert, she in a new Worth gown;
Alith and I were talking ovah on Bothton town
Of things intenth and thoulful. I begged her me love to cwown.
But Alith, alath, wath stubborn, ovah in Bothton town,
She'd be a bwothah to me, she thaid, but wouldn't be Mitheth Bwown.

In Defiance of the Kaiser

As Recited by Robert Irving Bush, the International Reader

Three years before the beginning of the Great War the Kaiser and his immediate associates held an evening party at the Royal Palace in Berlin. The American Ambassador was one of the guests. It was a comparatively small party, about twenty-five in all. The guests were gathered in front of the great fireplace after the elaborate banquet. From beyond the masses of tropical plants which masked the apartment where the orchestra was concealed came the exquisite strains of a Russian air, played on the violin by a master hand.

(Soft violin music heard, very pp. A phonograph record may be used if a soft enough tone can be produced.)

With that intense love of music that is so marked a characteristic of educated Germans, all present stopped conversation and listened, with every evidence of pleasure, to the solo. As the last notes died away the storm of applause broke out with spontaneous enthusiasm.

"Exquisite!" "What a master touch!" "What splendid technique!" "Wundershoen!" were expressions heard on all sides. The Kaiser was delighted. "It is a long time since I have heard anything that approached such a brilliant performance. I sometimes wonder how the lower classes can cultivate such tastes!"

The Crown Prince threw a malicious glance at

the American Ambassador as he said: "The lower orders have some things that we don't possess by birthright of nobility."

The fact that every person in the room, except the Ambassador, had a title made the remark and sneer doubly suggestive. But the Ambassador was equal to the occasion: "True, your Highness; brains, for example."

"Let us have a solo," said the Kaiser. "One solo—bid the player give us Deutschland Ueber Alles." He raised his finger. "Tell the man who played to come into the salon. I desire him to play before my guests." A servant retired.

A minute later he returned, and hesitatingly approached his master. The Kaiser looked up. "Well?"

The servant said something in a low voice.

"What!" The Kaiser arose. "Say that I command it."

The words were uttered too loudly not to attract attention, and inquiring looks were directed toward the Emperor. He was annoyed. Accustomed to instant and unquestioning obedience all his life, he would brook no excuses. "What do you think, friends? This great genius, whose playing you did him the honor to admire, actually has the insolence to send me a reply begging to be excused!"

A chorus of laughter followed. The idea was too absurd; the servant had not understood; the man was insane. There was a blunder somewhere, and the Kaiser was not accustomed to tolerate blunders.

The servant entered again. This time he was in terror. The man fairly shook.

"Come here!" said the Emperor.

There was a painful silence in the great apartment. The lackey slowly approached. Again he whispered in a low voice, and then stepped quickly back, as if fearful of a blow.

The Kaiser looked up. His face was livid with rage. His eyes fairly blazed with anger. "What! You dog! This message to ME! Rudolph, follow this fellow and drag that player here by force—by force, do you hear!" He controlled his anger a moment later and turned to his guests and said: "Your pardon, my friends, but what do you think the dog had the temerity to reply to my summons: 'I am not the servant of the Kaiser, nor a citizen of this country. I shall not play Deutschland Ueber Alles!' By Heaven, we shall see. I shall make the dog play here before you until you bid him go, and then he shall be flogged before he is flung from my doors."

The guests said nothing. They waited, with bated breath and painful interest, the ending of the scene. Men stood with compressed lips; ladies sat in nervous apprehension; the Emperor walked up and down like an enraged lion.

There was a sound outside—a scuffle—some confusion. All eyes were directed toward the great door. The culprit was dragged in, and from every lip came the simultaneous exclamation: "A woman!"

A young girl, bearing a violin and a bow, appeared struggling with the lackeys. With a quick gesture she threw them aside and advanced to the center of the room.

"Who is responsible for this outrage—this assault?"

"Silence!" the Kaiser thundered; "you were commanded here by me."

"Commanded? I am a free-born woman, a citizen of the United States. Who commands ME?"

"Silence, girl!" It was the Crown Prince who spoke. "You are in the presence of his Imperial Majesty, the Kaiser! It is his command that you play Deutschland Ueber Alles for his guests."

"That I shall never do!"

The Emperor was beside himself with rage. He pulled a dog-whip from the wall and advanced toward the girl. She never moved. There was no sign of fear; simply defiance.

"Take hold of her arm and compel the bow to cross the strings! What will save you now?"

"This will!" The answer of the girl, shrill and clear as a trumpet sound, rang out, as with a quick motion she raised the violin far above her head, and before any one could prevent, dashed her beloved instrument into a thousand pieces at the feet of the Kaiser.

He was baffled—beaten. There was a gasp as he made a motion with the whip. The American Ambassador sprang forward to avert the blow, but it was needless. The whip fell from the Kaiser's hands and he fell forward. The strain had been too much, and the next moment he was unconscious.

Confusion followed. The Ambassador sprang to the side of the girl. "Quick, come with me," he said in a low voice, and led her from the room while the guests and the servants thronged around their prostrate master. A cab was called and the girl fled from the city, safe at last! The Kaiser had been defied.

"Uncle Tom's Cabin" at the Op'ry House

A Song-Monologue

(This selection may be given by male or female. Special costuming is not necessary, but it adds to the effect. The words are to be sung in a rather rapid tempo, but slow enough for the audience to comprehend the points of the last lines of each stanza, technically known as "punch" lines. Play no interlude between the stanzas, but run them all together without a pause. The performer may act

as her own accompanist in this number. NOTE: *These verses are the sole property of Walter Ben Hare, and are inserted in this book for the sole use of amateur readers. This selection is protected by copyright and may not be used on the vaudeville stage or on Chautauqua circuits without a signed permit from the author. Violation of this warning will be severely dealt with by the law.*)

Way last spring, I think in April,
  A show troup landed in our town;
They give an op'ry called Uncle Tom's Cabin,
  'Twas better than a cirkis with a ring-tail clown.

A coal-black colored man lived in a cabin,
  With his wife and children, all snug and nice,
When along come a pesky oversee-er
  And chased Elizy acrost the ice.

Settin' on the bank was a great big bloodhound,
  Lizy kicked him outa the way—
I thought right there there was goin' to be a murder
  But the gosh-dinged critter was stuffed with hay.

Next scene out came little Evey
  And old Uncle Tom with a wreath on his head;
I knowed right away that the child wasn't healthy,
  So I wasn't surprised when she dropped down dead.

Curtain came down; I was feelin' mighty sorry
  Fer the little dead gal with her curls and her smile—
Bunkoed ag'in, by Gosh! there I saw her,
  A-sellin' her photygraphs down the aisle.

Sis Hopkins and Her Beau, Bilious

COSTUME.—Boy's shoes, striped stockings (these may be white stockings striped with red grease paint), ill-fitting calico dress longer behind than before and buttoned down the front with white bone buttons. Hair slicked straight back and tied in pigtails that curve out behind the ears, the curve being made by pieces of wire hidden in the hair. Funny little round straw hat, with an elastic under the chin. Red bandana handkerchief neatly folded in a square and pinned to the front of the dress with a safety-pin. Walk and stand pigeon-toed and in walking it is good business to occasionally catch the right toe on the left heel. This selection is obviously a burlesque and must be played in a broad comedy fashion.

(SIS *sticks her head in the door; the head only is visible to the audience.*) Say, ain't it my turn next? All right, I'm comin'. (*Enter.*) Here I be, sassy as a woodchuck and twicet as handsome. You know who I be, don't you? I'm little Sis Hopkins from Skinny-marink Crossroads down in Toadhunter Holler. (*Leans over and speaks confidentially to the audience.*) You know I'm the only one in this here show who's a real actress. The others is only ham-chewers, but I've had experience with a real show troup. I led one of the bloodhounds in Uncle Tom's Cabin street pee-rade last time it played down in the Skinny-marink opery house. Them other gals out there is all jellix of me.

You know why they're jellix? 'Cause I'm better lookin' than what they is, and I got a beau. (*Giggles.*) You orter see my beau. His name's Bilious Buttonbuster, and he's a great big fat boy, weighs purt' nigh a hundred an' eighty pounds, an' only four foot high. Took me in to see the side-show last summer when the cirkis come to town, an' I snum! ef they didn't try to keep him there to be the fat boy. Great big feller, Bilious is, weighs purt' nigh a hundred an' eighty-nine pounds, an' ivery inch of him is love.

The feller in the cirkis had me sing a song fer him. I told him I'd had my voice brought out by a singin' teacher, an' he said I'd orter have it sent back an' kep' on cold storage. But my singin' teacher said I had an awful fine voice; she said it was so mellow. I told Maw that the teacher said I had a mellow voice, and Maw said: "That's right, mellow means rotten." But Bilious jest loves my voice. He says it allers makes him homesick, 'cause it reminds him of the hogs and things back on the farm. (*Laughs.*) Bilious is jest too cute to live. You'd hardly expect sech a fat feller to be so cute, but he is. And he's jes' as fat as he is cute. Weighs purt' nigh two hundred pounds. (*Pause; then ecstatically.*) And ivery inch of him is love.

The folks that got up this show said that I was to sing you-uns a song. I'm goin' to do it. It's a real pathetic song; folks allers cries, er sump'm, ivery time I sing it. Sometimes they git so wrought up they git up and go out to relieve their emotions. The song I'm going to render fer you is a love song. (*Giggles.*) It's Biliouses favorite. He gits so sentimental when I sing it, and do you know—a fat man when he gits sentimental is sump'm awful—and Bilious is a great big feller. He ain't so big up and down, but I tell you he's a whopper round the middle.

I don't reckon any of you-uns iver heerd this song, as it was especially wrote fer me by ——— (*insert local name*), and no one else has ever had the nerve to sing it. My execution is sump'm wonderful—you'll all be in favor of it when you hear me. I sing this song real pathetic in parts—and then in other parts I'm jest as playful as a kitten. When I come to the pathetic parts, you'll know it, 'cause I make gestures there.

I allers think of Bilious when I'm singin' pathetic. It's the kind he loves the most. The name of the song is "My Bonnie Lies Over the Ocean," and sometimes when I git to singin' it real good, I actually git sea-sick, 'cause the way I sing you kin jist see the ocean and the bonnie and iverything. Jist imagine the bonnie's name is Bilious and he weighs purt' nigh two hundred and fifteen pounds. Now the piano will jest gimme a chord in B flat minor and I'll make my bow. (*Chord—bows awkwardly.*) Ain't that graceful? (*Sings nasally and somewhat off the key, but not too much.*)

My bonnie lies over the ocean,
                      (*Gestures with* R. *hand.*)
  My bonnie lies over the sea, (*With* L. *hand.*)
My bonnie lies over the ocean, (*Both hands.*)
  Oh, bring back my bonnie to me. (*Entreaty.*)

(*Speaks.*) Ain't that sad? I'm expressing the lacerated feelings of a gal whose tender heart has been deserted by her bonnie. You kin read the emotions in my face. (*Sings rapidly.*)

  Bring back, bring back,
   Oh, bring back my bonnie to me, to me;
  Bring back, bring back,
   Oh-ooo, bring back my bonnie to me!

(*Speaks.*) The second spasm is even more emotional than the first. Ain't none of the big emotional singers kin git ahead of me. I kin emosh jist as good as ary one of 'em. This second part is the one that Bilious likes the best. Sometimes tears come in his eyes when I sing it, and he feels the pathos of it a-throbbin' all through his two hundred and fifty pounds. (*Sings.*)

O, blow, ye winds, over the ocean, (*Gesture front.*)
 O, blow, ye winds, over the sea, (*Gesture to* L.)
O, blow, ye winds, over the ocean, (*Upward sweep.*)
 And bring back my bonnie to to me.
       (*Downward sweep.*)
(*Fast.*) Bring back, bring back,
   Oh, bring back my bonnie to me, to me;
  Bring back, bring back,
(*Drawl.*) Oh-o-o-o, bring back my bonnie to me.

(*Speaks.*) Wait till you hear the third verse, and git out your handkerchief if you've got a tear left in your body.

  Last night as I lay on my pillow,
       (*Pantomime sleep.*)
  Last night as I lay on my bed, (*Gesture.*)
  Last night as I lay on my pillow, (*Weeps.*)

Oh, oh! I dunno whether I kin go on er not. I git so emotional. (*Sings.*)

I dreamed that my bonnie was dead.

(*Sing chorus rapidly and make an awkward bow.*) That's where they ginerally applause me.

Bilious took me into town once to a high-toned fashionable dance. He wore a full dress suit, an' purt' nigh busted out of it. The hull upper part of his vest was busted anyway, and he had long tails on his coat. (*Laughs.*) Never see sich a sight in all my life, and him a-weighing purt' nigh three hundred pounds. (*Whispers.*) And ivery inch of him love! I wore a peekaboo waist at the party, all trimmed down the front with mayonnaise. It was a real swell party and Bilious was the swellest one there. But they certainly did have the skimpiest refreshments I ever saw in my life. Nothin' but two little skinny sandwiches and a coupla green plums. Bilious said they was olivers; tasted like salt mackerel to me. After we et we went into the ballroom. Iverybody was dancin', and some was writin' on little bits of paper who they was to dance with. Feller come up to me and said, "Say, Sis, is your program full?" I was never so mortified in all my life. "What's that?" says I. "I asked you was your program full," says he. I says, "Look here, you city dude, it takes more'n two skinny sandwiches and a coupla olivers to fill MY program, and don't you forgit it! (*Exit indignantly.*)

Whoa There, January!

(A Comedy Monologue concerning a prim old maid who drove the colt January to a glorious victory.)

AS RECITED BY EMMA DEE RANDLE, THE CHAUTAUQUA STAR

Miss Priscilla Parks was the primmest old maid in the county. You know the kind I mean, hair parted right slap smack in the middle, with jest as many locks on one side as on t'other, kinda old-fashioned in her dress, and her back allers held up so straight you'd think it was primed up with a poker instid of a backbone. She was a good-hearted lady, Miss Priscilla was, and right popular in the village—but, as I say, awfully strict. Any man that drunk a little, or chawed tobaccer or used bad language, or even smoked a pipe—was headed plum down the primrose path, accordin' to Miss Priscilla. So it wasn't much wonder that she was an old maid. And she hated hoss-racin' worser'n poison. Kinda funny, too, when her paw was the horse-raciest man in the county and was known far and wide for the speed of his colts.

Squire Parks was a good professin' church member too, but nothin' at all as strict as Miss Priscilla. Fact is, I never seen *sich* a strict Presbyterian as she was. But she had a heart overflowing with kindness and was a master hand in caring for the sick and needy. She gave 'em jelly and soup and roses and kind words, as well as some useful advice about livin' up to the ten commandments and keeping the Sabbath.

The squire had a colt named January, and I TELL YOU he was some high-stepper. One Saturday the squire hitched up his team to the family buggy and drove off to town. He calculated on returning on Monday. Miss Priscilla got all ready for church on Sunday morning and was real set back to find the team and buggy missing. But she had a will of her own, and told Tim Parker, the hired man, to hitch up January to the open wagon, so as she wouldn't be late for Sunday-school, the meeting-house being more than five miles away from the Parkses' farm. Miss Priscilla had never driven January before and she eyed him anxiously, asking Tim Parker if the colt had any " go " in him at all. Tim calculated that he had—a leetle—but he didn't tell her that the squire had been training January every morning for the past month, that he was the speediest colt in the whole county and could burn up the road like blue-greased lightning. But it wouldn't have scared Miss Priscilla if she'd known the truth, for she understood a thing or two about driving a colt herself. She was old Squire Parkses' daughter even if she was dead set agin' hoss-racin' and sich.

She started off and January went along jest as nice and smooth as you please. Just a leetle skittish he was, but Miss Priscilla thought that was due to his youth and the natural spirits of a bright April morning. When January found out that he wasn't due for his usual morning gallop he subsided and fell into a slow trot, much to Miss Priscilla's peace of mind, for fast driving on the Sabbath was a thing she never *could* and never *would* tolerate. They jogged along for about two miles when all of a sudden Miss Priscilla heard some one behind her coming down the hill lickety-split. She looked around. It was young Jim Dawson drivin' his blooded mare. January heard the rapid patter patter of the feet of the mare and stretched out his neck a leetle and felt of the bit. He threw up his head and quickened his pace.

Miss Priscilla pulled on the reins and said, "Whoa there, January!" January pricked up his ears and snorted. The blooded mare was closin' up behind 'em. The colt knew his own mettle; he could easy outdistance the mare—would he do it? —dare he do it? Closer and closer came the patter patter of the mare. She was going to pass 'em. "Whoa there, January; whoa, I say!" The mare was right behind him. Coming like a sky-rocket. Patter, patter—closer, closer! Will he let her pass? No, *no*, NO! A country mare going to pass *him?* Never! He bore down on the bit and threw himself forward in the harness until the traces nearly burst.

(*Imitate driving.*) "Whoa there, January! Whoa, old fellow!" (*Triumphantly.*) But January would not whoa! He put his ears straight ahead of him, stuck out his nose, planted his hind feet down with a thud, thud—threw out his forefeet like piston rods—and was off like the wind. Young Jim Dawson scented a race. The nose of his mare was just at the end of Miss Priscilla's wagon. (*Loud.*) Jim Dawson touched his mare with his whip and she speeded up. They were almost neck and neck.

(*More didactic.*) It was a clear road and a fair

start. Miss Priscilla would have given Jim Dawson a look that would have withered him dead, only she didn't dasst to take her eyes off'n January. (*Through clenched teeth, with driving gestures.*) She braced her feet, gathered up the reins in both hands and pulled for dear life. The mare (*Terror*) forged ahead and Jim Dawson let out a wild laugh of exultation, a sneer of defiance. Whoo! (*Increasing tempo.*) That laugh started every ounce of Miss Priscilla's sporting blood right into her face. (*Bold, loud, fast.*) She was the daughter of old Squire Parks, the raciest man in the county. She *forgot* her principles, she *forgot* it was the Sabbath, she forgot she was headed for the meeting-house, *she forgot she was a Presbyterian.* (*Loud, with a touch of pathos.*) She only knew that Jim Dawson was laughing at her, sneering at her, and that his mare was beating her colt. (*Louder, tense.*) She leaned over the dashboard, her bunnet knocked down over one eye. (*Louder.*) She didn't yell "Whoa there, January!" now. (*Yell.*) She yelled, "Go THERE, JANUARY! Go, GO!" And January flew.

(*Calmer.*) He stretched his nose a little farther out and broke into a gallop. (*More excited.*) His nose was at Jim Dawson's buggy. (*Loud.*) Miss Priscilla screamed, "Go there, January, go like the wind!" And January went like the whirlwind. He passes the buggy. The hosses are neck and neck. Go on, January, show your blood; go on—go on, for Miss Priscilla—go! Inch by inch the colt forged ahead. He passes the mare like a bird on the wing. Miss Priscilla gives a scream of triumph. "I beat him, I beat him! Whee! I beat him!"

Jim Dawson slowed down, but January had his blood up and raced on like a flash o' lightning. Down the hill and through the wood and over the bridge. The wagon swayed and Miss Priscilla clutched the reins for life. The meeting-house was only half a mile away. (*Faster, a touch of terror.*) There was a string of buggies ahead of 'em on the road. January shook his head, got the bit where he wanted it, stuck out his nose and *flew!*

Miss Priscilla tried to yell "Whoa!" but the road was so rough she could only gurgle and gargle at the rate she was racin'. The colt sped on and the wagon bumped after him like a cork in the river. Up and down—up and down. Miss Priscilla met the wagon seat half-way each time it bounced. Her bunnet flew off and hung rakishly down her back, her shawl broke loose and streamed like a flag of victory behind her. "Whoa there, January, whoa!" Bump, bump! "Whoa, I say, you pesky critter!" Bump, bump! (*Loud, exasperated.*) "Whoa, darn ye, whoa!"

On sped the colt. Deacon Jones, in his old rig, was just ahead of 'em. He tried to turn out, but it was no use. Blooey! The Deacon was whirled around just as Miss Priscilla lost her balance and tumbled over backward into the body of the wagon, waving both feet riotously in the air. Rigs drew aside and the people flocked out of the meeting-house and crowded the porch to see Miss Priscilla go by. And she *did* go by! Whir! bang! clatter! whiz! She *flew* by.

Three miles past the church January decided that it was time to stop, and he stood by the roadside and panted and panted as Miss Priscilla climbed out of the wagon and adjusted herself to her clothing as well as she could. She put on her bunnet and shawl and looked all around her. (*Hands up in horror.*) She thought of the crowd that had been in front of the meeting-house, and her appearance as she passed them, flat on her back with the sole of each foot pointing straight at the sky. She glanced reproachfully at January as she climbed into the wagon.

"My sakes alive, January, ain't you ashamed of yourself? It's a providential mercy that I still got a breath of life left in my body. Hoss-racin' on a Sunday! My land of love. (*Snicker.*) I must have made a pretty spectacle of myself before all the brethren and sisters of the church, but, thank goodness, I had my new stockings on! Git up there, January!"

Deceitful Man

A Study in Impersonation

Shakespeare tells us seven ages
   Constitute the life of man;
Reckoning from the earliest stages
   Till he ends the mortal span.
Woman always is an angel,
   One can always believe her—
But a *Man* through seven ages
   Is a base deceiver.

As soon as he can lisp he says:
   "Bobby didn't det it,
I 'ist left the cover off
   The jelly; pussy et it.
Honest, Ma, don't det d' switch,
   What's d' use of dat?
I's been des' as dood as dold,
   It musta been the cat!"

Hear him when a little older:
   "Honest, hope to die,
Gotta go to grandma's funeral,
   It—just—makes—me cry!
Boss, I gotta leave three thirty,
   Won't be back till dark!"
Lights a cigarette and hies him
   To the baseball park,

DECEITFUL MAN

For the third scene hear the lover:
  "Gee, Bess, you're some girl;
You're the first and you're the only
  Kid that makes me whirl.
I'll be true to you forever,
  Married life is bliss—
Bessie, you're the only girl I
  Ever tried to kiss!"

Hubby stumbles home at morning,
  At a wild night's end,
"Honest, darling, I've been sitting
  Up with my sick friend.
He's a Mason, I'm a Mason,
  It's Masonic law,
Not to leave him till the daybreak;
  Don't be angry, Ma!"

Fifth scene, goes with step determined
  His son to interview,
And lies: "My boy"—kerswat! kerswish!
  "This hurts *me* more than you!"
He next has leisure on his hands
  And fills a can with bait;
He hooks a minnow, then he swears
  Ten pounds to be its weight.

Last age, see him as a grandpa,
  A tear is in his eye,
"Johnny, though I'm nearly seventy,
  I've never told a lie!"
Woman always is an angel,
  One can always believe her;
But a man through seven ages
  Is a base deceiver.

Old King Faro's Daughter

Farce-Monologue in Negro Dialect (Georgia)

AS RECITED BY PROF. W. E. VAUGHAN, OF
MEMPHIS, TENN.

All you chilluns gather up here now close to youah Sunday School teacher 'kase I'm gwine to expound de story ob old King Faro's daughter. Liza Williams, you stop scratchin' yourself an' listen to de Scriptures I'm expoundin'. Lily Bud Dawkins, you wake up Perfumery Perkins and make her stay awake. Now all pay strict attention kase it's my endeavor to lead you all outa youah sinful ways and make you walk into de straight an' narrow path of de Baptis' faith. Dis here lesson is from de Old Testament book ob Moses.

Long time ago way down in de lan' ob Egyp', dey lib an ole king what was name Faro, count'n him bein' a bad man an' a crap-shootin' gambler. And he had a daughter what was naturally called Miss Faro. Now dis here Miss Faro was a high-steppin', high-brown 'Gyptian princess, count'n her paw bein' a king, and she neber step out in de street ner nowhars without her han'maiden a-followin' her an' a-totin' her trail. She lib in a li'l palace all by herself wif nobody cept'n her han'maiden. An' it didn't make no diffunce what Miss Faro was a-doin', whether she was a-washin' or a-ironin' or a-scrubbin' de flo', she always wore de purtiest silk and satin dresses you ebber laid youah eyes on.

OLD KING FARO'S DAUGHTER

One mawnin' Miss Faro was a-sottin' out in de front porch in a big rockin'-chair, a-rockin' back-'ards an' for'ards and singin' (*sing*), "Run along, Li'l Mary, run along, Li'l Mary"; her han'maiden was a-standin' by her side shooin' de flies off'n her with a big turkey-struck fan. All at once Miss Faro stop a-rockin' and drawed herself up in all her majestical heights an' turn round to de han'maiden an' say, kinda haughty-like, kase you know she was a king's daughter, an' say, "Han'maiden!"

Han'maiden she say, "Whut you want, Miss Faro?"

Miss Faro she say, "I gwine 'dopt me a chile."

Han'maiden she say, "Say you is, Miss Faro?"

Miss Faro she say, "I mos' certainly is. You go up-stairs an' fotch me dat peachblow dress wif all dat passamentery work an' all dat grill work up an' down de front, kase dat's how I'm gwine 'proach de king."

Han'maiden fotch de peachblow dress down an' Miss Faro 'rayed herself mos' scan'lously. She takes her hair outa de shucks an' tied it up wif a blue ribbon, she pinned on all her dimonses an' emeralds an' rhubarbs, an' den she get a whole lot ob dis Hoyt's perfume an' fumigated herself from top to toe. Den she started fo' de king's palace wif de han'maiden walkin' behin' her totin' her trail.

Well, when she got to de king's palace dat mawnin', de king he was holdin' a consultation wif his co't. Dere he was a-settin' on his throne, wif his long plush gyarments reachin' clean to de flo'. De consulary standin' on his right-han' side and de vice-araries on his lef'-han' side, a skepter in his right han', a diadem in his lef' han', clown settin' on top ob his haid, rubies an' sapphires an' topazes strung all up and down his shirt bosom an' diamonses was scattered 'round on de flo' so thick, ef you'd a-been barefoot, cut yo' feet all to pieces.

Well, Miss Faro walks along de co't, han'maiden followin' long 'hind her, totin' her trail, and at last she gits up to de big door, and she stood there and shuck de wrinkles outa her dress an' goes into de great hall where de king's sittin' at, an' jes' marched up de aisle wif her shoulders throwed back an' de han'maiden followin' 'long 'hind her, totin' her trail. She walked up pas' all de soldiers an' de co't wif her nose stuck up in de air an' her chist throwed out—didn't look to de lef'-han' side ner de right-han' side, kase she was de king's daughter. When she gits up in front ob de throne she flop herself right down in de middle ob de flo' in front ob de king. Den she look up at him kinda cute-like an' say, "Good-mawnin', poppy."

He look down at her and say, "Good-mawnin', honey, what yo' want dis mawnin'?"

She say, "Poppy, I wants to 'dopt me a chile."

"Well, go on an' 'dopt you a chile den, what yo' come 'round here botherin' me for? I's a-holdin' consultations wif ma vice-araries an' ma consularies dis mawnin', dog gawn dey skin, dey don' pay no 'tention to me—but dey got to listen to me dis mawnin', er I'm gwine a bust 'em wide open."

She say, "Poppy, I wants to 'vize wif you, kase I'm skeerd I might 'dopt some ob dese poor white trash."

He say, "You git outa here an' quit botherin' me."

Den she riz up and *histed her carcass to de do'*, han'maiden followin' long 'hind her, totin' her trail.

When she got on de outside dere was de chillun ob Israel strewed along on bofe sides ob de road what old King Herod done slew, 'count'n him see de Star ob Bethlehem risin' in de East. Den she met up wif old Brudder Norah, what built de Ark an' de 'Postle Peter what was a-walkin' on de water ob de Sea ob Galilee.

She say, "Brudder Norah an' 'Postle Peter, good-mawnin'."

And dey say good-mawnin' to her back an' bow mighty low, kase she was a king's daughter, han'-maiden followin' long 'hind her, totin' her trail.

An' she say, "I's gwine 'dopt me a chile. Kin you-all tell me whar I kin find me a nice high-brown pickaninny to 'dopt?"

Den Brudder Norah, he say, "You go on down to de River Euphrates an' look aroun' and you'll see what you'll see! Yas'm, Miss Faro, look 'round right sharp, an' you'll see what you'll see!"

Miss Faro she thank him and she went on a walkin' wif her nose in de air an' she shoulder throwed back jes' like a nigga soldier come back from France. Han'maiden followin' long 'hind her, totin' her trail.

Pretty soon dey comes to de River Euphrates wot water de lan' ob Egypt. Den Miss Faro she heerd a cur'ous kin' ob a noise ober whar de river was at. She raise up her skirt to keep de dus' off her ankles an' started through de cane-brake, wid de han'maiden followin' long 'hind her, totin' her trail. Pretty soon dey come to de aidge ob de water, an' what you suppose she see floatin' on de bosom ob de tide? A bullrush!

Now a bullrush am one ob dese yere contraptions 'bout so long, an' so wide an' so high an' lined inside wif tar an' feathers to keep de water from oozin' through. When Miss Faro sees de bullrush she drawed herself up in all her majestical heights an' say to de han'maiden:

"Han'maiden, I command you to wade out in dat water an' fotch me dat bullrush."

Han'maiden she say: "Miss Faro, I don' wanta be wadin' out in dat water an' hab de bull-frogs jumpin' off'n de logs at me an' dem snaikes a-wroppin' round my laigs an' d' mud a-oozing 'tween ma toes."

Miss Faro she say: "Han'maiden, ef you don' wade out in dat water an' fotch me dat bullrush, I'm gwinter bus' you wide open."

Now han'maiden she done been busted wide open once er twicet 'fore by Miss Faro an' she ain' honin' for no mo' bustin's. So she wade out in de water an' rech down, pick up de bullrush an' fotch it to Miss Faro, an' laid de burden down at her feet. Miss Faro she reach down, open de bullrush, an'—what you s'pose she foun' in dere?

Naw, 'twasn't no snaik, Perfumery Perkins, how come a snaik git in dat bullrush? It was a baby! Purtiest li'l baby you ever laid yo' eyes upon. When Miss Faro seed d' baby she turn roun' to d' han'maiden and say, "Han'maiden, dis is de chile ob my adoption. (*Intone, imitating the "shouting" of a negro when under religious excitement.*) Oh, oh, dis am de chile! Oh, oh, foun' him in de bullrush! Oh, oh, I believe I'm gwine to shout. I's about to go off in a transom."

Han'maiden she say to Miss Faro, "Miss Faro, I's been a-washin' an' a-ironin' all de mawnin' an' I'm so tired dat I can't tote you home ef you goes off in a transom. Restrain your religious convictions till yo' gets to de privacy ob youah apartments an' shout like de white folks."

So Miss Faro she 'strained her religious convictions till she git home, an' den she took de li'l baby an' washed it an' cleansed it an' combed its hair and put a purty white dress on it, den she walk out to whar 'Postle Peter sot a-fishin' wid five loaves an' two li'l fishes in de waters ob Galilee, han'maiden followin' 'long 'hind her, totin' her trail, an' she showed 'Postle Peter what she done found.

'Postle Peter he say, " What you gwine to name him, Miss Faro?"

Miss Faro she say: " I gwine name him Moses in recumembrance ob dat good ole man Moses what led de Chillun ob Israel outa dey bondage into de Red Sea, an' made a fatted calf out ob gold for de Prodigal Son what had done returned from eatin' manna in de wilderness. I gwine to name him Moses after dat man!"

An' 'Postle Peter led Miss Faro and li'l Moses down to de River Euphrates, han'maiden followin' 'long 'hind her, totin' her trail, an' he baptized li'l Moses in de good ole Baptist way, while Miss Faro went into anudder transom on de bank ob de river.

Sarah Jane

A Pianologue.

(*Play some light pretty air while reciting the first two stanzas.*)

The late lamented Sarah Jane,
    Alas how she is missed!
For two and twenty years she was
    Our village organist.
She studied music half her life,
    The other half she taught it,
But genius is a fickle flame,
    It seems she never caught it.

When Sarah Jane was but a child
    Her house was next to ours;
I used to love to hear her play,
    She'd practise hours and hours.   (*Imitate.*)
With do, de, mi—and mi, re, do,
    'Twas like a harp of gold—
To hear her practise all day long
    When she was six years old.

(*Play some simple air with one finger, striking false notes, etc. Note: do not prolong the agony.*)

When Sarah Jane was sweet sixteen,
    A dainty winsome maid,
She graduated out of High
    And on the program played.

SARAH JANE

She didn't choose a high-brow piece
   By Wagner or Gounod,
But played the tune that won my heart
   So many years ago.

(Repeat the same air, correctly.)

Then Sarah spent four years in town,
   'Twas just the same old story,
She won the medal of her class
   At the Conservatory.
And played like sixty, day and night—
   I heard her grand recital,
And this is how she played the piece—
   I knew it by its title.

(Play same selection with grand opera runs, cadenzas, etc.)

Sarah came back to the old home town,
   And lived her simple life,
As daughter, sister, sweetheart true,
   As fond and loving wife.
She taught and led the choir in church,
   At every dance and party,
She hammered out that same old tune,
   She played it a la cartey.

(Play same selection as a jazz dance piece.)

And so she lived; but when at last
   Our Sarah passed away,
Another organist came to me
   And asked me what to play.

She said, "What was her fav'rite tune?"
　　I gave it with a sigh,
And at the funeral we heard
　　The tune of days gone by.

(Play same selection as a dirge.)

The late lamented Sarah Jane,
(Play same air softly and correctly.)
　　Alas how she is missed,
For two and twenty years she was
　　Our village organist.
And now in lands beyond the skies
　　Her golden harp she plays,
While angel bands all join and sing
　　That song of other days.

Her First Ride in an Ottymobile

A Prize-Winning Farce-Monologue

[This monologue may be given in an old maid costume and make-up. If this is done have a wrap and comical bonnet ready to don at the proper time. A chair stands facing the audience. The back hair of the reader should be arranged in such a manner that it will come down when indicated in the text. If desired mechanical effects may be worked behind the scenes. An egg-beater when whirled gives a fair imitation of the throbbing engine, and an auto horn could be used advantageously.]

Land sakes, Aunt Jane, who's that just druv into our yard? It's a ottymobile; I kin see it standin' down there under the ellum tree. It's Joshuway, you say? Joshuway Slabb? In an ottymobile! Wall, wonders'll never cease. Tell him I'll be right down. Give him a pa'm leaf fan, set him out on the stoop, and tell him I'll be there in a minute. Yes, hurry up. (*Calls after her.*) Give him the almanac to read. (*Comes to the front of the platform, putting on hat, and primping at audience as in a mirror.*)

Joshuway's bought him a new ottymobile. (*Giggles.*) Wall, ef that hain't the limit. I think he'd have done much better ef he'd a bought a house and lot and a marriage license. (*Puts on wrap.*) He's been a-sparkin' me fer nigh onto thirteen years and has never got up his courage to propose yet. Comes twicet a week and once on Sundays. But in spite of everything I kin do he's the bashfullest critter in Splinter Township. And now he's got a ottymobile. (*Puts on gloves.*) Wall, mebbe he means business at last. (*Giggles.*) Mebbe it's fer my weddin' present. Must have cost a mint o' money, too. It'll make a lovely weddin' present. But, lawsy me, I certainly hope he kin git up enough gumption to propose. There, I'm all ready. I suppose he wants me to go drivin' with him down Lover's Lane. (*Giggles.*) Wall, I'm ready to go with him even if the road leads to the parson's.

(*Pretends to meet him.*) Land sakes, Joshuway, I wasn't expectin' you till to-night. Lookin' mighty spry, and all dressed up, too. How are you?

Oh, yes, I'm well. (*Turns and pretends to see the auto.*) Fer the land sakes, Joshuway Slabb, where *did* you git that ottymobile? You bought it? Well, of all things! Ain't it nice? Kin you drive it? You *kin?* They showed you how to down in the village? Well, well, well! Wonders'll never cease. (*Pause, listening to him.*) Why, of course, I'll be tickled to death to go ridin' with you. I jest love a ottymobile. Land o' Goshen, I've been watchin' them things whizz by my door fer the last six er seven years, but I never calculated I'd git to ride in one. My, ain't it purty? All black and shiney and spick span new, ain't it? And it's got upholsteried leather and seats to set down on and a top and everything. Air you sure it's right safe, Joshuway?

I seen them things whizz right down that road

HER FIRST RIDE IN AN OTTYMOBILE 71

there like a cheriot of fire. (*Laughs.*) Sometimes they went so quick it 'ud take two persons to see 'em go by, one to say, "Here she comes" and t'other to say (*With a burst of laughter*), "There she goes!"

Oh, no, I ain't afeerd as long as you're with me, Joshuway. I got every confidence in you, but fer the land sakes, don't lose control of the critter. Holt tight onto the reins—that's all I ast, Joshuway, jest holt tight onto the reins. How did you ever larn to drive it? (*Pause.*) You've took three lessons. My, you're sech a masterful man, you kin jest do *anything* when you git your mind made up to it. (*Sigh.*) But sometimes it takes a powerful long time fer you to make up your mind.

Where do you want me to set? Up in front er there in the back? In front? (*Pause.*) I'll git more air in front? All right, I'm willin'. Now hist me in, keerful now, and fer the land sakes don't let it git to goin' till I git good and sot. (*Pretend to climb in, sit in chair on the platform.*) My, ain't it comfortable? (*Leans back.*) I feel so stylish settin' right up here on the front seat of a real ottymobile, jest fer all the world like I was a queen livin' in the Walled-off Castoria Hotel in New York City. (*Sees imaginary steering wheel.*) What's that little wheel fer, Joshuway? (*Pause.*) Well, I want to know. That's what you drive it with? It looked to me like the rudder of a one-hoss steamship. (*Burst of laughter on last two words.*)

Yes, I'm all ready, let's go. Git in. I'm jest a dyin' to see how the critter runs. (*Anxiously.*) What you doing? Crankin' it? What do you have to do that fer? To make it start? Well, land o' Goshen, this here ottymobile reminds me of an old jinny mule we used to have named Lily. Sometimes she'd balk and stand jist as still as a broomstick, till brother Hiram twisted her tail. That's the only thing 'at 'ud start her. (*Laughs.*) And now you've got to twist the ottymobile's tail jist fer all the world like our old (*burst of laughter*) mule Jinny!

(*Noise of a whirling egg-beater heard behind the scenes.*) There you got her. (*Excited.*) She's a-goin'. She's a-throbbin' like sixty. Joshuway, *Joshuway*, git in quick and grab holt of the lines, she's liable to start down the hill. (*Pause, as he is supposed to climb in beside her.*) Oh, I hope I hain't crowdin' you, Joshuway. (*Nudges up to him.*) There ain't much room in here, and I'm a little bit skeerd, too. (*Looks up in his face coyly.*) But you won't let her run away, will you? Yes, I'm all ready. Let her go. (*Suddenly give a quick jerk backward as the car starts.*) Glory to Goshen, I like to swallered my teeth. The pesky critter rared right up on its hind legs like you'd hit her with a whip. Now, go slow, Joshuway. Go slow at first anyway, till I git used to the sensation. There, we're goin'—we're goin'. Down the hill. Not too fast, Joshuway. Oh, Josh-u-way, pull on the reins. Holt her back.

(*Smiles.*) My, ain't it nice? (*Sees some one approaching down* R. *aisle.*) There comes the Bolivars in their spring wagon. I hope they'll see us. Turn out, Joshuway, turn out—they're headin' right toward us. (*Honk of horn is heard behind the scene.*) Turn out! (*Pause; anxious action.*) Not so fast! Glory to Goshen, we whizzed by 'em like a shot outa a gun. I thought you'd run plum smack into 'em. Look out, *look out!* There's a

HER FIRST RIDE IN AN OTTYMOBILE 73

cow. (*Honk, honk!*) Joshuway, you mustn't be so reckless. My land, you almost hit her. Sim Bolivar will never know how close he come to havin' minced cow fer supper.

Now let's not go so fast up the hill. That's right, slow her down. (*Jolt.*) My, ain't it rocky? (*Sudden bump; angry tone.*) Joshuway Slabb, you look out now. There's a puddle. Josh-u-way! (*Screams.*) Ohhh! That muddy water jest splattered all over me. Make her go slower. (*Arranges hair.*) I must look a sight. Slow down a leetle, anyway. I want to see the scenery an' things.

Now we're comin' to the Sniderses' farm. Spurt her up a little bit and let's drive by there in style. (*Egg-beater is twirled slower.*) It's a-slowin' down. Let's go faster. You can't? Why not? Hit her with the whip. It's a-slowin', it's a-stoppin'. It's stopped. (*End egg-beater noise.*) It ain't throbbin' no more. (*Suddenly, in fright.*) Oh, Joshuway, we're a-goin' backwards. We're a-slidin' down hill. Help, help! We'll tumble in the river. Stop her, Joshuway, stop her! (*Screams and throws her arms around his neck.*) Oh, I'm real fainty. I'm a-goin' to faint. Josh-u-way. There, she's stopped.

Joshuway Slabb, you take your arms right away! How dasst you? Huggin' me right out here in the public highway, and we ain't even engaged. (*Pause.*) It ain't your fault that we're not engaged? My glory to Goshen, whose fault is it? (*Change tone to bashful.*) Oh, Joshuway, I never dreamed that you was in love with me. (*Pause.*) My answer? Oh, this is so sudden. (*Pause.*) Yes, I know you've been a-sparkin' me fer thirteen years, but it's sudden jist the same. (*Pause, then assume a meek voice.*) Yes, Joshuway, I will. (*Sudden scream.*) Oh, Joshuway! Joshuway Slabb, you mustn't! You mustn't! (*Indignant tone changes to submission.*) Suppose somebody 'ud see us. Well, jest one then. (*Pause, business of being kissed.*) You're sich a masterful man. (*Sudden.*) Lemme be! Somebody's comin'. There comes a wagon. Start up the machine again. That's right, crank it harder. (*Egg-beater noise.*) There it's a-goin'. Git in, it's goin' to start. (*Resumes her seat.*) Here we go. (*Speaks to some one passing.*) Evening, Sister Hendricks. (*To Joshuway.*) My, it's jest going grand. Whip her up a little as we go past the Sniderses. They're out in the yard. Howdy, Almiry, howdy!

(*Looks around.*) My, ain't the scenery pretty here on the top of the hill? Slow down a little bit now, Joshuway. We're starting down hill. Make her go slower. You can't? You forget how. Oh, Joshuway, not so fast! (*Bump.*) My glory to Goshen! Look out, there's a big bump. (*Screams.*) Awww! (*Sudden flop in chair, hair comes down.*) Oh, somebody stop us. Help, help, we're a-runnin' away. Murder! Look out! You're runnin' into that tree. Turn in! Ohh! (*Sudden bump.*) O-h-h-h! (*Long groan.*) I'm killed. (*Egg-beater noise ends.*) I know I am. (*Faintly.*) Are you still alive, Joshuway? Joshuway, air you hurt? (*Runs to him.*) Speak to me, speak to me, one little word to your Alviny, your wife-to-be! Speak! (*Pleading tone changes to indignation.*) Joshuway Slabb, that was a swear word. Ain't you ashamed? Go back to the Sniderses and git their team and haul us home.

Here, I'll help you. Ouch, I got a crick in my back. Well, anyhow, we got engaged to be married, and that's sump'm, ain't it? (*Pause.*) When? (*Pause.*) Oh, some time next week. Step keerful now, Joshuway, step keerful. (*Hobbles out.*)

Hiram's Blunder

It comes to light that Hiram Swett
    Went out last night to dinner
And pulled a breach of etiquette
    That was indeed a winner.
His poor wife claims she's quite upset
    Because of Hiram's manners,
And says, of all the boobs she's met
    He carries off the banners.
He sneezed and choked in one fell swoop
    And filled the guests with terror,
And dropped his false teeth in the soup—
    Which was a social error!

Things We See on the Stage

A Musical Monologue

I went last night to Broadway
    To see the latest show,
I saw the movies, vaudeville,
    The opera stars aglow;
I saw the social drama,
    The musical comedy,
The scenery, tights, electric lights,
    That make a Folly Show!

(Put on large hat with burlesque paradise plumes, imitate musical comedy star, speaking the following lines to waltz music. If possible have a spotlight thrown on you.)

    There's a waltz, there's a waltz,
        And a dashing huzzar,
    And a sweet peasant maid,
        Neath the light of a star;
    And the chorus, so fair,
        Every shape, every age,
    And that is light opera
        As seen on the stage.

(If the reader is able to sing she might introduce one or two popular choruses from light opera at this point.)

Lightest kind of music,
  Show girls, turkey-trot,
Scenery, tights, electric lights,
  Everything but plot.

(*End music, remove hat.*)

Then to the Metropolitan,
  The high-brow opera home,
I heard the foreign warblers,
  From o'er the ocean's foam,
The hero, aged sixty,
  Was vocalizing dago,
Soprano weighed three hundred pounds,
  And troubled with lumbago.
Hefty tenor warbled,
  Fat sopranos screech,
Critics say it's perfect,
  Seats ten dollars each.

(*Reader dons large sombrero and false mustache, stands before spot-light and sings in deep voice burlesquing " Duet from Il Trovatore."*)

  Ah, I have sighed to rest-eo,
  On the soft side of a plank-eo,
  I cannot sleep,
  Some one will weep,
  I feel like da sheep,
  I feel like da sheep,
  I cannot weep.

(*Hurriedly throw off the hat and mustache, put on picture hat again, without missing a note, sing in high soprano.*)

Ah, I hear you sleeping,
I will not gorgetta,
Hurry, let us finish,
And eat da spagetta,
    Ah, ah, ah!
Now let us finish
And eat-a spagett.

(Throw off picture hat and come down C. *again.)*

Here we have the drama,
    Loving couple wrangle,
Villain comes, handsome chap,
    See the old triangle!
Wife elopes with villain,
    Husband says his say,
Wife repents, joy intense,
    Matinee to-day!

(Throw shawl over head, stand in spot-light. Music: "Flower Song.")

Violets, violets, who'll buy my fresh roasted violets? Here I am out in the snow-storm. Out in the blinding snow with the wind tearing around me. *(Paper snow is thrown over her from sides.)* No home, no mother—only baby and I. I'm alone, alone on the streets of New York. Ha, what do I see? Reginald McSweeney approaches in his limousine. *(Honk, honk! heard behind scenes.)* Still do I defy him. I may be poor, but Elaine the Beautiful Flower Girl remains as spotless as the driven snow.

(Throw off shawl, put on hat and mustache.)

Ah, ha, 'tis she! At last I have found you. Long have I waited for this minute, and now at last you are in me power. Hist, hist! What was that? Who said hist? Was it you or the audience? You recognize me, Josephine, do you not? Yes, stare at me, shiver if you will, for I am Reginald McSweeney, the wolf in sheep's clothing. Bah, bah! Do not kneel to me. Bah! I said I was in sheep's clothing. In spite of your disguise, well do I know thee, Josephine Lemon!

(*Throw off hat and mustache, put on shawl.*)

I am no longer a lemon, I am married. Four years ago, innocent cheild that I was, I married a traveling man. But where he is now I know not. He kept on traveling. But I know I am in your power. Have mercy! Have you not a spark of mercy in your icey breast? What have I done that you should treat a poor girl in this manner? No, no! Release me! (*Weakly.*) Help, help! You'd bind me to the tracks. (*Kneels.*) Have you no mercy? Have you no heart? Oh, do not do this horrible thing. Help, help! (*Prone on floor.*) He's bound me to the railroad track and the eleven forty-two is due in two minutes. Oh, I shall be ruined forever. He's gone! I am about to perish. But no! (*Sits up.*) I forgot! These are the B. and O. tracks (*Substitute local railroad's name*) and the B. and O. is always five hours late. (*Totters to front.*) I am shaved, I am shaved!

> Villains, vamps and heroines,
>   Every size and age,
> That is melodrama,
>   Seen upon the stage!

Nora Has Her Picture Took

FARCE-MONOLOGUE IN IRISH DIALECT

[This selection may be given in costume, the reader wearing a very gaudy dress liberally trimmed in bright green, a large hat of pink and a red or yellow parasol. She carries a huge shopping bag, made of pasteboard covered with dark cloth and bits of bright thread to simulate beadwork.]

Mary Ann Finnigan, y'll never guess what I've been doin' this day. Small wonder that I'm all drissed up in me bist, wid Mrs. Marvin's new Paris hat on me head, herself bein' gone away to spind the wake ind in the country. (*Triumphantly.*) I've been havin' me picture took. (*Pause.*) Oh, ho—y' thought something terrible had happened to me, did y'? Sure, it *was* terrible. I wouldn't go through the like of it agin fer twinty photygraphs. (*Sighs.*) I nearly died.

I had me picture took for O'Gilly, me finansay. I've been intendin' to do it fer some time, but I niver got the nerve until this afternoon off, me havin' little or nawthin' to do at the house, herself havin' took the twins wid her to the country. She wanted me to go wid 'em, but country life is entirely too fat-i-gooing for me, so I tould her I had heart palpitations and she gave me a vacation till Monday marnin'. But be that as it may.

I wint around on the Avenoo to the swellest photograph gallery in the city. I was determined to do the thing up in style. The man met me at the door as nice as you plaze and relieved me of foive dollars in advance and thin he showed me into the studio, and my, my! just as soon as he opened the door the divil a sich a smill I iver smilt at home or in Ireland as was in that room. And right before me very face stood the operator. Ah, Mary Ann, I nearly died.

I was fairly pink wid fright at what stood before me.

"Where did that thing come from?" says I.

"Hush," says the man who let me in, "that's old Mr. Chromo, the operator." And wid that he went out in the reception room and lift me alone wid ould Mr. Chromo. An ould skinny man he was, wid a white face and wan foot in the grave. Sure, after what happened to me I wish now I'd pushed the other foot in along wid the first. Oh me, oh my, sich an old scarecrow. Mary Ann, he looked like a hundred years old, an' ninety-nine of 'em spint in the hospital. Y' niver saw sich a crature, starved-lookin' he was, and fer all the world like wan of thim skillitons in the doctor's office. He come toward me. Oh, I nearly died.

"Come this way," says ould Mr. Chromo, and he planted me doon in a big chair forninst a bit of a box histed up on three legs wid two eye-holes in the front of it. He pushed it, he did, and straightened it and whaled it around, until it was pinted right at me like a gun, or a' cannon, er wan of thim death-dealin' instruments of the Dootch. Thin up to me he comes and takes me two shoulders and straightens me round in the chair. And thin—what d' ye think!—he clapped a grapplin' iron to the back of me, stuck me head in it, clinched it tight, till I could nather move to the right ner to the left.

"Now, don't move," says he, "kape very still until I come back," and away he goes to the little dark room beyant.

There I sot wid me head in a vice. Begorry, I was scared green and the precipitation begun to pour outa me face like a hail-storm. It went through me mind all of a sudden, Mary Ann Finnigan, that the two of thim were up to some divilmint in the room beyant—mayhap they was going to kidnap me and hould me for ransom, like I've sane it done in the movie pitchers. Me wid me gould watch and chain an' me bead bag an' the missus' hat on—belike they thought I was wan of thim Astorbilt heiresses. The more I thought about it the more excited I got. The man who let me in had a black mustache and iviry wan who goes to the movie pitchers knows that that's the sign of a villain, and the other wan, oh me, oh my, that ould sickly lookin' skilliton would niver stop short of murder or assassination. And all the time the little black box was pointed straight at me. So I out of me sate and around to the back of the box to satisfy meself that there was no murderous weepons concaled there under the black rag. But jist as I put me hands onto the cloth, may all the Saints in Hivin presarve me, but there stood the ould bag o' bones be the side o' me, grabbin' me wrist. I nearly died!

"What are you doin' *here?*" says he. "Didn't I tell you to kape your position and not stir?"

"Sure," says I, "I was jist goin' to look through the little windys at meself sated over there in the chair, to see how I look sittin' down. That's all I was doin'."

"Don't be frightened," says he, "there's nothin' goin' to hurt y'. To quiet you I'll allow you to look through the camera to see how a picture is taken."

And he sated himself in the chair. I paked through to see what koind of a pitcher I was goin' to make, but what I saw nearly gave me the high-stiricks, thin and there.

"Is that me?" says I.

"Av course not, it's meself," says he, "that's the way you'll look when I'm takin' the picture."

"Are you sure av it?" says I.

"I am," says he.

"That I'll look that way?" says I.

"Exactly, that identical way," says he.

Thin I let out a yell and started to duck fer the door, for would you belave me, Mary Ann Finnigan, there he was standin' in front of me as plain as you plaze, wid his *heels* in the air and his head on the floor. I nearly died.

"Give me back me money," says I, "and lave me out of here at once. It's a respictable girl I am, and niver stood on me head yet for any man, alive or dead!"

And I was that excited that I rushed at him and grabbed him wid wan hand and hit him over the head wid the missus' new bead bag wid the other.

"Stand me on me head, is it?" yells I, wid another whack over his back wid the hand-bag.

"Here, here, what's going on here?" said the wan wid the black mustache running in from the other room.

"What do y' think he was going to do to me?" says I. "That ould chromo was going to set me down there and put iron grappers on me back till I could nather move nor scream, thin he was goin' to stand me on the top of me head and maybe murder me intirely."

Both of the two of 'em nearly boosted themselves wid laughing. My, my, but I was exasperated. I nearly died! Finally the black mustache explained to me that I wouldn't be standin' on me head at all, at all,—that was only the picther, you understand, and me settin' there in the chair right side up all through the performance.

"Come," says he, "I'll fix y' meself."

Thin he lead me to the chair and sated me as polite as a Frinchman. The ould chromo wint about his business and I was glad he did for the likes of him I niver saw before ather on the earth or in the waters beneath the earth, and the Saints know I niver want to say it again. The big man twisted me head around the same way the ould skilliton did and clamped it in a vice.

"What are y' doin' that for?" says I to him.

"Be aisy," says himself, "and kape still the way I fix you. I don't want the whole of your face to appear in the picture."

"You don't want the hole in me face to appear in the pitcher? (*Angrily.*) What do ye mane by the hole in me face? I'll have ye understand that I have *no* holes in me face. Me face is unholy, so it is. Hole in me face, indade! Say, what do ye think me face is anyhow, a pace of Swiss cheese?" Was I mad? I nearly died!

"No, no," says he, "I mane I don't want all of your face in the picture."

"Will, me fi-nansay Mike O'Gilly wants all of me face, and I want all of me face, so what you want is not particular at all. What do ye think I want? Jist a half a picture whin I paid me good money fer a whole wan? Do ye think I want a picture of wan ear, er wan eye, er wan nose only? Indade not! It's me whole face I'll have, er none at all." "No, no," says he, "I'm going to snap your profile."

Now, Mary Ann Finnigan, I'm a lady, and I was brought up like a lady, and me mother was a lady before me, but right thin and there I forgot me ladylike qualifications. Up I bounced again and shook me beaded hand-bag under his nose, me heart palpitatin' wid angry emotions.

"Snap me profile?" yells I, "sure, I'd like to see you snap me profile, er any other part of me anatomy! I didn't come here to be snapped. I'm a paceful woman, I am, and a lady, but if ye git frish wid me, I'll hand ye wan wid me fist, so I will."

He gave a long sigh and apologized like a gintleman. I wint back, ca'm an' dignified, and sated meself in the chair loike a quane upon her throne. He arranged me face, stuck his head under the black cloth, tould me to smile like a happy little bird, and I did—(*smiles*) something wint click-click, and me picture was took. But I wouldn't go through the like of it agin fer twinty dollars. My, my, whin I think of that ould chromo and the way I was trated, I nearly **died!**

The Musicale

A Study in Expression

Mr. Boggs isn't much of a musician, but he is worth something over a couple of million and Mrs. Boggs, anxious to hold her slight footing in society, decided to give a musicale. The palace is crowded, all the world has come accompanied by its wife. A string sextet is wrestling valiantly with the Sextet from Lucia on the improvised stage. (*Soft music "Sextet from Lucia," on phonograph.*) Mrs. Gusher, who just *adores* music, has cornered Boggs, the unhappy host.

".Oh, Mr. Boggs, don't you positively adore the sextet?"

Boggs tried to appear at his ease, as if chatting with Mrs. Gusher about the unknown sextet was a thing he did every day.

"Adore it? Wall, I can't say. To tell y' the truth, Mis' Gusher, I never rid in one. Is it a towering-car er a limousine? Mine's a ninety-hoss power ——"

He was interrupted by an unknown gentleman on his left.

"Say, these musicales are torture, aren't they?"

Boggs admitted they were.

"I never heard of this Boggs until my wife told me I had to bring her to-night. Know him?"

Mr. Boggs began to feel uncomfortable, decidedly uncomfortable.

"Well, er, I——" But he was interrupted by the unknown.

"Say, old man, who's that freak with the ptomaine-poison expression on her face over by the door?" Boggs looked at the freak and smiled grimly to himself as he answered: "That's Mrs. Boggs. She's giving the musicale, you know." Mrs. Gusher, tired of Boggs's neglect, pulled his arm and said sweetly: "Do you play, Mr. Boggs?"

Boggs felt right at home immediately. "A little poker, Mis' Gusher; ever set in yourself?"

Mrs. Gusher was properly horrified and replied frigidly: "I was about to ask if you played the Barcarolle from Hoffmann?"

Boggs was stumped. "Nope," he answered cheerfully, "I play pinochle accordin' to Hoyle, though." Mrs. Gusher turned away and Boggs turned to his unknown friend on his left. Just then an attenuated strawberry-blonde began a violin solo.

"Sufferin' fish-hooks!" said Boggs.

"What say?" said the unknown.

"Look what's fiddlin' up there for us. Where on earth did Maria find that human torch? Of all the jokes on the human anatomy——"

The unknown gentleman interrupted with some asperity, "That, sir, is my daughter. May I ask who you are?"

It was Boggs's inning. "Oh, never mind me. I'm only the husband of the freak with the ptomaine-poison expression." And the stranger made a hasty exit.

A Quiet Man at a Baseball Game

(*This may be given by a man as a monologue, substituting the personal pronouns for the name Smithers.*)

Mr. Smithers is a mild-mannered gentleman who attends strictly to business. He is polite and honest and speaks in a gentle, well modulated voice. Every evening he is home with his family. He plays with the children, discusses wall paper and measles and radishes with his wife and neighbors, takes up the collection in the Episcopal church, and attends to his business in a mild, quiet, dignified manner during his working hours. In fact, he is an unobtrusive, plodding, meek and worthy citizen.

See him on a summer afternoon. It is three forty-five and he is dictating to his stenographer.

"I think we have time for one more letter, Miss Casey. If you please. (*Dictates.*) The Goldstein Wholesale Plumbing Company, Kennebec, Kentucky. Dear Sir: Your letter of even date received. Would say that I can fill your order at once. Have filed bonds to guarantee contract and will have banker wire you to-night. I appreciate the fact that thirty-two thousand dollars is my net profit for the deal and would——!"

The clock struck three. "Er—Miss Casey, never mind the rest of it now. Let it wait until to-morrow. (*Pause.*) Yes, I know there's a chance of losing the contract, but I have an im-

A QUIET MAN AT A BASEBALL GAME 89

portant engagement in twenty minutes. (*Pause.*) Oh, yes, it is very important. There are some things much more important than a thirty-two thousand dollar bonus. Good-afternoon."

Fifteen minutes later he is in the bleachers at the ball game. There are three men on bases and the man at the bat has had two strikes called on him. Mr. Smithers rises up in his bleacher seat, emits a war-whoop, jumps up and down and lets out a choice collection of yells and phrases that his business associates would never suspect was in him.

"Hitterout, Spike, man, hitterout; that rummy in the box won't let a run walk in, he's going to pass you a puzzle and you gotta hit it out—you GOTTA hit it out. Watch his eye! Watch his EYE! WATCH his eye! (*Pause breathlessly.*) He hitit, he hitit! Right out in the left pasture. Beat it, boy, beat it—run, you clam—RUN—slide, SLIDE—aw, why don't you slide, you bonehead crawfish. Wha's a matter with yer feet? RUN, don't crawl—good boy, slide, SLIDE, SLIDE!

Good! He's safe—he's—WHAT? OUT! Thief! You robber! You, you, YOU ——!"

Mr. Smithers is plainly disgusted and he shows his disgust by turning to his right hand neighbor and slapping him in the neck.

"Who's up? Eh? Casey? Aw thunder, he's a dub. He couldn't hit a biplane. He's got double astigmatism in both arms. He's—E—YOW! Yow, yow, YOW! Lookit, lookit! Run, Casey, you green lobster, run—RUN—safe on first—E—YOW—E!"

And Mr. Smithers in his frenzy stamps his heel on the tender and invalid toes of the gentleman on his left.

"Who's up now? Who? Hannigan? Great!

some kid, Hannigan. Hooray—e—yay—lookit, lookit, Spike's at the third doily, Casey's stole second—Jack muffed it and Hannigan's safe on—what? Thief! Highwayman! Assassin! Robber!—robber! Oh, he's safe? Oh! Sure, I said he was safe. He's safe. Wow!"

And Mr. Smithers, meek, unobtrusive Mr. Smithers, mashes both his fists through the new spring lid of the party in front of him. But nobody minds, they are fans, fans, and they are watching the grrreat American game of baseball.

Different

Encore Verse

When a love-sick little dove
Wants to see his lady love
In a rather informal way,
He simply says Coo-ee,
And she replies Coo-ee,
   In a manner recherché.
There's no need of chaperones
When he coos in dulcet tones,
His sweet little soft Coo-ee;
And they flutter to a nest,
Dovey Cupid does the rest—
But it's different with you and me.

Once a little Teddy bear,
Met a lady bear, so fair;
   She ensnared him in an ursine way.
He queried, "Will you wed?"
She drooped her coy head
And murmured, "Si'l vous plais."
Then he winked his button eye,
And she heaved a little sigh—
A sweet little soft, Ah me!
Then they danced a wedding rag
To a syncopated drag!
But it's different with you and me.

Mrs. Gilhooley's Bungaloo

Have yeez heard about me good fortune? Ould Mrs. Pinnymint doied, she did, and what do ye think?—She lift me a furnished bungaloo in her will. Oi been her cook fer twinty years and she had so much money she didn't know who to lave it all to. She lift Dinnis the coachman a bungaloo, and she lift Nora the housemaid a bungaloo, and she lift me a furnished bungaloo. Foive rooms it had, and ivery wan of thim furnished.

Whin Oi heard of me good fortune Oi took the strate car and wint out to look at me bungaloo. Oh, it was a lovely house, all riddy fer me to move into, wid the pots and the pans and the sheets and the pillow-cases, and the chairs and tables and aven a pianny in the front room and a bird cage in the windy. And what do ye think Oi found in the cellar?

One dozen bottles of champagney water, all corked up toight and covered wid spider's webs. Oi knew well that Oi'd be arristed under the bone-dry law if they found me wid the champagney water in me possession and it nearly broke me heart to have to throw it away.

But law is law, so Oi sated meself in the kitchen, havin' pulled all the curtains doon, and decided to put timptation out of me path foriver. Oi pulled the cark from wan bottle and it exploded like dynamite, but Oi turned it upside down in the sink and emptied all its contints, all except a small wine-glass full which Oi drank.

Thin Oi pulled the cark from another bottle and it exploded like dynamite, but Oi turned it upside down in the sink and emptied all its contints, all except a small wine-glass full which Oi drank.

Oi thin pulled the cark from another bottle and it exploded like dynamite, but Oi emptied it upside down in the sink and turned all its contints into a small wine-glass full which Oi drank.

Oi thin pulled the cark from another dynamite, and it exploded like a bottle, but Oi emptied a small wine-glass full in the sink, and turned it upside down whin Oi drank.

Oi thin removed another sink—Oi mane—another bottle from the cark—and Oi emptied it upside down into the small wine-glass and it exploded like dynamite which Oi drank.

Oi thin pulled the sink from another cark—Oi mane Oi pulled the bottle from the dynamite—and it exploded loike a sink into a small wine-glass full which Oi drank.

Oi thin drankled another small copple—Oi mane Oi dunkled a tump from 'nother copple—you say, me frind, Oi mane Oi drank another small dynamite—Oi cackled—Oi mane Oi conckled—Oi mane the contints into the sink—or into the dynamite—Oi mane into me, Oi pulled the cark. Well, onyhow, Oi did it to all av thim twelve bottles.

Be this toime the bottles was all empty and Oi was full of dynamite. The kitchen range began to dance forward and backward, and Oi stiddied the sink wid wan hand and began to count the bottles wid the other. Be this toime they was whirlin' rapidly round me. Oi counted twenty-sivin out of the dozen.

Thin Oi decided that Oi have to clane the cobwebses off'n the bottles, but they was revolvin' round me loike a merry-go-round at Cooney Island. Oi sat on the flure and counted sixty-four av thim as they wint by.

But by pritindin' indifference to thim and springin' at thim as they wint by whin they was aff their guard, Oi managed to capture all of thim, by their necks. Oi placed thim beside me on the flure and they immadiately began bouncin' oop and down. Oi counted thim again as they performed their evolutions. Two of thim Oi hild in me two hands and Oi counted the others as they bounced. Would you belave it, at me final count there was ninety-five. Thin Oi wint to slape.

Lilian

AN ENCORE "SELL" THAT IS A NEVER-FAILING HIT

Airy Fairy Lilian her friends called her in that aristocratic old southern town where she was born and reared amid the palmettos and the pines of southern Louisiana, and Airy Fairy Lilian her friends called her at Oberlin College. When she was a Junior at college she fell in love—fell in love with a picture. It belonged to her roommate, Lucy Gates of Atlanta, and was a likeness of her brother Bob.

Lucy wrote to Bob and told him about her wonderful, charming room-mate who was so much taken with his picture. Bob, a successful young business man of Atlanta, was much impressed and one day when he was in a romantic mood he addressed a letter to Airy Fairy Lilian and enclosed it in a letter to his sister. Thus the romance began. Lilian sent him letters, an embroidered pillow-top and pound after pound of fudge—he, in return, sent her lilies, placing a standing order with an Ohio firm to send Lilian a large corsage every week.

The romance ripened into love and in her Senior year, Lilian and Bob became engaged. The wedding was to take place in July, a few weeks after Lilian returned home. All preparations had been made, the affair was to be in a large church, and Lilian was to have eight bridesmaids. Two days before the wedding Lilian received a telegram from Bob stating that owing to business pressure he would be unable to see her until the morning of their wedding day.

Her wedding day dawned bright and clear, the ceremony was to be at high noon, but at nine o'clock Lilian received another telegram. Bob would meet her in the chancel of the church immediately before the wedding. He had arrived and everything was lovely.

The church was crowded, the wedding march pealed forth and Lilian slowly marched down the aisle on her father's arm, preceded by her bridesmaids. At the same time Bob and his best man marched in from the chancel. They met at the altar. She raised her eyes and saw him for the first time in her life—a shudder!—a pause! Bob was a negro! Tall, young, athletic, but black—the pure-blood of his race proclaiming itself in every feature.

Lilian gazed at him, bowed her head and the minister began the service.

(*Voice in the audience.*) She married a negro?

Oh, yes. You see, Lilian was a full-blooded negro herself!

NOTE.—Have some one in the audience ready with the question at the proper time. Give the answer clearly and make a quick exit.

A Christmas Heroine

This selection may be given as a dialogue by two girls, or may easily be arranged as a reading.

CHARACTERS

Miss Mealy.—*A middle-aged lady in a winter house dress. Spectacles.*

Peggy.—*A little waif from the poorhouse. Ragged white stockings torn and patched with red and black. Large, old shoes. Short gray calico skirt, torn and patched. Old ragged white waist. Hair in curls.*

Scene.—*A boarding-house parlor. Table at* c. *with a long cloth.* Peggy *is concealed beneath the table. Enter* Miss Mealy.

Miss M. Well, this is indeed a merry Christmas. (*Sits.*) I've received several beautiful presents showing that my friends still remember me, even if I am forced to live in a third-rate boarding-house. (*Looks around.*)

Female Voice (*outside*). Peggy! Peg-gyyy!

Miss M. That Mis' Bacon is a reg'lar slave driver with that poor little kitchen girl. I saw her going to the grocery this morning in the blinding snow wearin' a ragged little dress and not a rubber overshoe er an umbrella to her back. And on Christmas day, too. Just because Mis' Bacon adopted her from an Orphan Asylum she thinks she has the right to work her to death. I know what

I'll do. I'll go down in the kitchen and give her an orange. (PEGGY *sticks her foot out from under the table and wriggles it.*) Heavens, it's a mouse. (*Jumps into a rocking-chair and has a difficult time trying to maintain her balance.*) Shoo, shoo! Oh, I'm frightened to death. It's a horrid, ugly mouse. Help, help!

PEG. (*sticks her head out from under table*). Naw, it hain't no mouse. It's only me.

MISS M. (*gets down*). Why, Peggy Malone, what on earth are you doing down there?

PEG. (*on floor*). Hidin'.

MISS M. But why are you hiding? Whom are you hiding from?

PEG. The old lady. She's got a grouch on. And I got a pain in me mitt. (*Holds up left hand, bound in rags.*) See! I'm all in. (*Pathetically.*) Oh, Miss Mealy, I just can't work no more. Me head's a buzzin' and I'm all ready to drop.

MISS M. (*sympathetically*). You've injured your hand?

PEG. Yas'm, I knows dat me own self.

MISS M. But how did it happen?

PEG. Burned it.

MISS M. Where? Where could you burn your hand?

PEG. All over me fingers and me wrist.

MISS M. No, no; I mean where was it burned?

PEG. Jest burned all over. Me fingers an' me wrist.

MISS M. Where were you when it burned?

PEG. Standin' right there with it. That's why it hurts.

MISS M. Peggy Malone, answer me. What burned your hand?

Peg. The fire.

Miss M. But where was the fire?

Peg. (*evasively*). I just stuck in me mitt, like that (*gesturing*), and it burned jest like paper. Kind o' foolish, wasn't I? Jest to go and stick me mitt in the fire. Sometimes I think I ain't got no sense at all. But it hurts sump'm awful.

Miss M. (*seated at* R.). Come here to me. Let me fix it for you.

Peg. (*with closed lips signifying "no"*). Um-um! Can't do it nohow. The old lady is after me. She's got a grouch on already, and she'd have 'leven different kinda cat fits if she finds out I burned me mitt.

Miss M. But it must be real painful.

Peg. Well, to tell the truth, it don't feel like no Sunday-school picnic, that's a cinch.

Miss M. You poor little thing. Here is some old linen. (*Takes it from hand-bag.*) Come here and let me bind it up for you.

Peg. (*as before*). Um-um. I'm skeerd.

Miss M. Why, Peggy, I won't hurt you.

Peg. Yes'm, I knows dat all right. I ain't skeerd of you. I'm skeerd of the old lady.

Miss M. But I won't tell her.

Peg. (*looking in her face*). Honest, won't you?

Miss M. (*pleasantly*). Honest, I won't.

Peg. Cross yer heart, an' hope you will never git married in all the world, if you tell?

Miss M. Yes. (*Crosses heart.*) There, are you satisfied now?

Peg. Yes, I'm satisfied. (*Crosses to her.*) There it is. (*Extends hand.*)

Miss M. Just let me take this rag off. (*Removes it.*)

Peg. Ouch! Gee whizz!

Miss M. Did I hurt you?

Peg. (*smiling*). No'm. Not much. I jest said that 'cause I was much obliged.

Miss M. Let me see. My, my, Peggy Malone, that's an awful burn. Why, you must have been crazy.

Peg. Dat's jest what I think meself. Honest, I don't believe I got a lick of sense. Burnin' me own hand like that. I must be a nut.

Miss M. Here, I'll put some vaseline on it.

Peg. Miss Mealy, you're an angel, a real fallen angel, fallen right off of a Christmas tree.

Miss M. Did you ever see a Christmas tree, Peg?

Peg. (*quickly*). You bet yer boots; I seen one last night. It was the biggest Christmas tree that ever was. It was all covered with——

Miss M. (*surprised*). Last night? Why, where were you last night?

Peg. (*looking at her a moment embarrassed, then laughing*). Did I say last night? That jest shows I'm clean batty. I meant last year. Honest, it was last year. Of course I never saw no Christmas tree last night. Ain't no Christmas trees up in my garret, is there? Ain't nothin' up there but rats, and mice, and beetles, and great big pincher bugs.

Miss M. Aren't you afraid of them?

Peg. Yes'm, of course I am. But it's me to the garret, or me to the orphunt asylum. An' a garret is better than a orphunt asylum, even if there is a hundred million rats and mice and pincher bugs.

Miss M. (*putting her arm around her*). You poor little thing. There, it's all bound up.

Peg. Honest, Miss Mealy, you're a kind lady,

that's what you are. To be helpin' the likes of me, when I ain't got sense enough to keep from burning me mitt up in the fire. You certainly are some doctor.

Miss M. Does it feel better, Peggy?

Peg. Does it? Why, it feels jest like it was eatin' a dish of strawberry ice-cream covered with chocolate. But it sure did hurt. Burning your mitt is enough to take the crimp out of any marcel that ever was waved.

Miss M. You have a merry disposition, Peggy, and a kind heart.

Peg. And say, you orter see me appetite. Me appetite is the best thing I got.

Miss M. Now you'd better go and tell Mrs. Bacon all about your accident.

Peg. Yes, and git sent back to the orphunt asylum. Not on yer life. I'm goin' to keep dark until me mitt gits well.

Voice (*outside*). Peg! Peg-gy!

Peg. You hear them musical strains? That means me. I gotta go and peel the onions fer the Christmas dinner. I jest love to peel onions, I don't think. (*Goes out.*)

Miss M. (*picks up newspaper*). Poor little thing, poor little homeless Peggy! (*Reads.*) " Fire at the Municipal Christmas Tree." (*Speaks.*) What's this? (*Reads.*) A terrible catastrophe was narrowly averted at the municipal Christmas tree exercises last night. The large hall was filled with children, and Mr. Alex. Byers, the Santa Claus of the evening, was handing out toy dolls when his whiskers caught fire. In a minute his whole costume was blazing. He pulled it off, but two little girls who had come forward for their dolls were a mass of flames. Suddenly a little ragged child jumped up on the platform, caught the two little girls, put out the fire, saved the lives of the children and ran out the back way before any one could catch their breath. No one knew the ragged little heroine. She escaped from the rear entrance with her poor little sleeve all ablaze. A collection was immediately taken up for the unknown heroine and contributions are solicited for the child. Steps have been taken to ascertain her identity. The child will be recommended for a Carnegie medal and over three hundred dollars have been subscribed for her benefit. (*Jumps up.*) It was Peggy! She tried to keep it from me for fear Mrs. Bacon would be angry. She ran away to go to the celebration, and now she's a heroine with over three hundred dollars waiting for her. Oh, I'm so glad. Peggy!

Enter PEGGY.

Miss M. Oh, Peggy, I've got the grandest news. Hurrah! there's the biggest Christmas present in the world waiting for you.

Peg. Fer me? Say, what you givin' me?

Miss M. It's all in the paper.

Peg. What is? Me Christmas present?

Miss M. Everything that happened last night.

Peg. Good-night; then dis is where I loses me job; it's back to the orphunt asylum fer me if Mis' Bacon ever finds out I was outa the house last night.

Miss M. You are a heroine; a little Christmas heroine. You're the bravest little girl in the whole town.

Peg. Aw, now, quit yer kiddin' me.

Miss M. Didn't you rescue two children last night at the risk of your own life?

Peg. Naw, I didn't do nothin'. All I did was to catch two kids and pull 'em outa danger. Dat's how I burned me mitt.

Miss M. You get your hat on and come with me. I'm going to take you right down to the newspaper office. They have over three hundred dollars for you.

Peg. Fer me? Me?

Miss M. Yes, for the little Christmas heroine.

Peg. I ain't no heroine. I dunno what you mean.

Miss M. Come with me to the office and you'll find out.

Peg. I can't do it. I gotta peel the onions fer dinner.

Miss M. Your days of peeling onions are over, Peggy Malone. I guess Mrs. Bacon will be a little kinder to you now. I'm going to tell her the whole story and then we'll all go and claim your reward.

Peg. If Mis' Bacon ever claims dat reward I'll never see a cent of it.

Miss M. Yes, you will. Leave all that to me. This is Christmas Day, Peggy Malone, come on and get your Christmas gift. (*They go out.*)

(*This dialogue is an arrangement of two scenes from Mr. Hare's popular play for girls, The Camp Fire Girls.*)

Bob's Girl

(Bob is a typical schoolboy of nine.)

I got a girl, an' I ain't ashamed—
  I love her with all my might,
And if you think you kin make fun o' her,
  Well, someone's got to fight.

She ain't got no long golden curls,
  She ain't got eyes of blue,
But she's got a smile and she's got a heart
  That's kind and lovin' too.

And when a feller's feelin' blue,
  You oughta see her smile,
And push yer shoulder an' pull yer hair,
  And talk like a friend fer a while.

I never took her to dancin' school,
  Ner to the baseball game,
I never sent her a valentine,
  But she's my girl, jest the same.

She's the finest girl in this here world,
  I never want no other;
She's my chum and my pal and my sweetheart, too,
  Who is she? She's my mother!

Tomboy

(For a girl of eight or nine.)

I don't care what the people say, I'd rather be a boy,
And howl and yell and run away, and fish and fight, oh joy!
I want to go for hick'ry nuts and join the baseball team,
Play shinny in the alley, and swipe some kid's ice-cream.
It's awful hard to be a girl and always be dressed up,
I'd rather be a boy, I would, er else a yaller pup.

One day I swiped my brother's clothes an' shoes an' Sunday hat,
An' dirtied up my face, I did, an' throwed stones at the cat,
An' went down where the circus was, away acrost the town,
An' walked around as big as life an' tried to pinch the clown.
Some kids was stealin' in the show behind the water tank,
I tried it, too, but I got caught, an' got an awful spank.

The teacher says she never seen a girl as bad as me;
One day she came to our house and she had to stay to tea;

I sneaked down to the drug-store, got a nickel's
   worth of Wheeze,
It's a whitish kind of powder that'll make folks
   sneeze and sneeze.
I did it just to entertain the teacher, you can bet,
And it entertained the whole of us; I ain't through
   laughin' yet.

First Pa began to sneeze so loud we thought he had
   the flu,
The teacher says, "Atchee, atchee!" an' Ma she
   says, "Kerchoo!"
Then Grandpa sneezed and said he thought "it
   really was the weather,"
The hired girl sneezed and brother sneezed, then all
   kerchooed together.
The puppy sneezed, and baby sneezed just like he
   had the croup,
And Grandma sneezed her false teeth out; they fell
   in teacher's soup.

Then Pa he took me by the ear and led me to the
   door,
And did a lot of other things that made me good
   and sore.
I think I'll run away and be a pirate chief or king—
But I'm a girl, and girls can't do a single naughty
   thing.
Wish I could yell an' howl an' run, an' fish an' fight,
   oh joy!
I don't care what the people say, I wisht I was a
   boy!

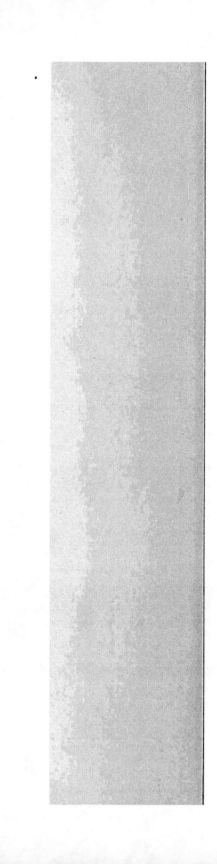

**University of California Library
Los Angeles**

This book is DUE on the last date stamped below.